Geospatial Technologies and Society

SCIENCES

Geography and Demography, Field Director – Denise Pumain

Cartography, Subject Head – Colette Cauvin-Reymond

Geospatial Technologies and Society

Uses, Imaginaries, and Critical Issues

Coordinated by
Thierry Joliveau
Matthieu Noucher

ISTE

WILEY

First published 2026 in Great Britain and the United States by ISTE Ltd and John Wiley & Sons, Inc.

ISTE Ltd
27-37 St George's Road
London SW19 4EU
UK

www.iste.co.uk

John Wiley & Sons, Inc.
111 River Street
Hoboken, NJ 07030
USA

www.wiley.com

Library of Congress Control Number: 2026934921

British Library Cataloguing-in-Publication Data
A CIP record for this book is available from the British Library
ISBN 978-1-78945-241-9

ERC code:
SH7 Human Mobility, Environment, and Space
SH7_10 GIS, spatial analysis; digital geography

The manufacturer's authorized representative according to the EU General Product Safety Regulation is Wiley-VCH GmbH, Boschstr. 12, 69469 Weinheim, Germany, e-mail: Product_Safety@wiley.com.

Contents

Foreword

Colette CAUVIN-REYMOND
CNRS, UMR LIVE, Strasbourg, France

As part of the ISTE SCIENCES collection, and more specifically the Geography and Demography field, there is a set of four books devoted to cartography. Cartography is both a scientific and artistic discipline (Robinson 1952, 1953) and is essential for any person, organization or institution that needs to process and represent geographic data to highlight its spatial characteristics. With current technological transformations, the continual increase in the use of Web 2.0 and the emergence of social networks, cartography is once again undergoing profound changes.

Over time, this discipline has undergone numerous mutations of varying degrees of significance, both conceptually and technically, the two being closely linked. As Rosnay (2008) puts it, "scientific progress and technological progress feed off each other". While its convergence with statistics at the end of the 19th century introduced new processes for translating localized phenomena, the advent of the computer and the subsequent birth of computerized cartography can be considered a revolution. This revolution was described in an article by Tobler in 1959, which announced both the stages of constructing an automated map and the basic principles of what would become Geographic Information Systems (GIS).

Between then and the early 1980s, this "new" cartography developed in two directions: on the one hand, computer cartography, which reproduced what had previously been done manually, and on the other hand, computer-assisted cartography, which opened up new avenues for creating innovative representations

Geospatial Technologies and Society,
coordinated by Thierry JOLIVEAU and Matthieu NOUCHER. © ISTE Ltd 2026.

and introducing methods for surface analysis[1]. In the early 1980s, microcomputers appeared, offering immense possibilities that were now available to a large number of people with minimal knowledge. "Making" maps seemed easier, updates became simpler and software of varying quality proliferated. The availability of the Internet, and especially the World Wide Web in 1994, further simplified the diffusion of graphic documents, which multiplied.

Alongside these new features, a fundamental transformation has taken place in the paradigm of communication, which is essential to the discipline. Despite automation and the advent of the Internet and the Web, the logic of a movement from author/producer (the one who designs and represents) to reader/user, sometimes with feedback, remains predominant. The reader, the one who looks at the map and uses it, remains passive, even if, thanks to animation and, above all, interactivity, they can move the document around, zoom in, fly over it, etc., but they cannot modify it. However, with the arrival of Web 2.0, the Semantic Web, and the advent of the Geoweb from 2005 to 2010 and especially 2015, a new turning point was reached: the reader/user, whoever they may be, became active. They can modify maps on the Web and even create their own. This marks the beginning of a new era in cartography, which will be explored in depth in the works intended for the ISTE Encyclopedia to present an up-to-date overview of this science and its transformations for the period 2010–2025.

The date of 2010 was chosen as the threshold because many cartography publications, both French[2] and English[3], were published in previous years. These publications outlined the characteristics of the discipline just before the "real" launch of the Semantic Web and highlighted the steps required to produce a map and the successive choices that had to be made. In fact, the construction of a map requires at least three main steps, each of which has a specific role. The first step is to build a localized database based on the geographical information provided. The second step involves processing and transforming the data in the database, focusing either on locations, thematic data or both simultaneously, taking time into account or not. This step results in the determination of the mode of representation adopted. The third step, communication and diffusion, relies not only on technical solutions, semiotic choices and the adoption of specific actions but also on knowledge of elements of perception and cognition.

1 This approach largely corresponds to the analytical cartography introduced by Tobler (1976, 2000).

2 Béguin and Pumain (1994); Cauvin et al. (2007, 2008, 2010).

3 Slocum et al. (2009).

These three steps[4] are interdependent and relate both to the issue at hand and to the relationship between the recipient/user of the map. They guided the selection of the four works chosen for the ISTE Encyclopedia, as they reflect the state of cartographic science at a time just prior to significant changes. This makes it easier to highlight, describe and characterize the specific and original features of the years 2010–2025. One book is devoted to each of these stages, and a fourth places cartography in its historical context, facilitating a better understanding of changes over time. Each book can be read independently of the others and in any order the reader wishes, according to their preferences and expectations. The same applies to each of the chapters.

Looking Anew at the History of Maps (De Biaggi and Vagnon-Chureau (forthcoming) aims to "outline the history of cartography as it stands today and as it continues to evolve, offering a summary of ideas and changes over the past 40 years to present a history of cartography that takes into account current thinking and research and, above all, opens up new avenues for exploration".

This book entitled *Geospatial Technologies and Society* focuses on localized data, its sources, characteristics and use in all areas involving spatial representations, at a time when digital technologies are gradually but fundamentally reshaping contemporary societies. The acquisition of spatially referenced data is no longer an activity reserved for specialists. As the sources of this geolocated data become more diverse, citizens themselves directly and often voluntarily provide geographic information (GPS surveys, cell phones, connected mobile objects, digital platforms, etc.); networking is immediate. The search for geographic information has therefore been profoundly transformed with the "emergence of the geonumerical mapping" (Joliveau 2007), revealing at the same time major challenges for the digital society.

Handling and Mapping Geographic Information (Cunty 2024) has a double objective. The first is to present methods and techniques for processing and transforming information to produce maps that respond specifically to each problem. These procedures take into account the spatial, thematic and temporal components of this information, and focus on spatial and spatiotemporal processes using indicators and models. The second objective is to highlight the importance of the interdependence of the different stages of cartographic construction, in which the processing stage is central. On the one hand, the latter must take into account the data to be processed, selected at the previous stage. On the other hand, it specifies, for the next stage, the original cartographic representations that will result from the

4 A similar pattern can be found in the latest edition of Robinson's reference work from 1995.

selected processing. These two objectives are systematically concretized with examples that provide a clear understanding of the methods and their contribution to our comprehension of the phenomenon under study.

Cartographic Communication (Mericskay 2024), meanwhile, focuses on the transmission of the map, as it is not a document that is produced to remain in one's pocket. It must become visible, and even potentially audible and tangible, by making use of graphic and other variables, possible actions and the properties of display devices. It is certainly in this volume that it will be easiest to identify the specific features of the current period, with the role of choices at all stages, the intertwining of these stages and collaborative work.

Using the same tools and resources, specialists and non-specialists will produce maps using what is available to them, but with different training, which creates risks. Indeed, the new cartographic processes, or now geovisualization processes, that are constantly emerging certainly have many positive aspects: ease of use and participation, diversity, attractiveness, etc. But like any new development, their advantages have a downside; each of these processes has specific properties that it is important to know, at least minimally, to apply them properly and avoid mistakes. The four books presented here show that we are moving toward new logics, new ways of conceiving, translating, processing and transcribing geographic information cartographically, but without necessarily rejecting what already existed. The world of maps, now the world of geodigitization and geovisualization, is no exception to the general evolution in which we are all participating, with our wealth and our worries. May these books help everyone to be aware of what they are producing and to use all these wonderful tools with discernment.

February 2026

References

Béguin, M. and Pumain, D. (1994). *La représentation des données géographiques. Statistique et cartographie*. Armand Colin, Paris.

Cauvin, C., Escobar, F., Serradj, A. (2010a). *Thematic Cartography and Transformations*, Volume 1. ISTE Ltd, London and Wiley, Hoboken, NJ.

Cauvin, C., Escobar, F., Serradj, A. (2010b). *Cartography and the Impact of the Quantitative Revolution*, Volume 2. ISTE Ltd, London and Wiley, Hoboken, NJ.

Cauvin, C., Escobar, F., Serradj, A. (2010c). *New Approaches in Thematic Cartography*, Volume 3. ISTE Ltd, London and Wiley, Hoboken, NJ.

Cunty, C. (ed.) (2024). *Handling and Mapping Geographic Information*. ISTE Ltd, London and Wiley, Hoboken, NJ.

De Biaggi, E., Vagnon-Chureau, E. (eds) (forthcoming) *Looking Anew at the History of Maps*. ISTE Ltd, London and Wiley, Hoboken, NJ.

Joliveau, T. (2007). Géomatique et géonumérisation. Monde géonumérique [Online]. Available at: https://mondegeonumerique.wordpress.com/geomatique-et-cie/geomatique-et-geonumerisation/.

Mericskay, B. (ed) (2024). *Cartographic Communication: Graphic Semiology, Semiotics and Geovisualization*. ISTE Ltd, London and Wiley, Hoboken, NJ.

Robinson, A.H. (1952). *The Look of Maps. An Examination of Cartographic Design*. University of Wisconsin, Madison, US.

Robinson, A.H. (1953). *Elements of Cartography*, 1st edition. John Wiley & Sons, New York.

Robinson, A.H., Morrison, J., Muehrcke, P.C., Kimerling, A.J., Gutpill, S.C. (1995). *Elements of Cartography*, 6th edition. John Wiley & Sons, New York.

Rosnay, (de), J. (2008). *2020 : les scenarios du futur. Comprendre le monde qui vient*. Éditions Fayard, Paris.

Slocum, T.A., McMaster, R.B., Kessler, F.C., Howard, H.H. (2009). *Thematic Cartography and Geography Visualization*, 3rd edition. Pearson Prentice Hall, Upper Saddle River, NJ.

Tobler, W.R. (1959). Automation and cartography. *The Geographical Review*, 49(4), 526–534.

Tobler, W.R. (1976). Analytical cartography. *The American Cartographer*, 3, 21–31.

Tobler, W.R. (2000). The development of analytical cartography. *Cartography and Geographic Information Science*, 27(3), 189–194.

Introduction

The Spatial Dimension of Digital Transformation

Thierry JOLIVEAU[1] **and Matthieu NOUCHER**[2]

[1]*UMR Environnement-Ville-Société, Université Jean Monnet de Saint-Etienne, France*

[2]*CNRS, UMR PASSAGES, Pessac, France*

The history of using computers to process spatial data is now quite long. The first steps were taken in the 1950s with mechanical machines, the ancestors of computers (Joliveau 2004). This was followed by the first applications in theoretical and quantitative geography, the conceptualization of Geographic Information Systems in the 1960s, the first software programs and scientific work in the 1980s, the structuring of a specific field of research and a professional sector in the 1990s, the systematization of the geoweb and the use of GPS in the 2000s, the spread of these technologies to the general public through microcomputers, then smartphones and the ubiquitous Internet in the 2010s, and the emergence of artificial intelligence in the 2020s. This accelerated overview of the decades is, of course, somewhat misleading, as it combines and articulates very different levels of analysis: scientific, technical, social, economic, cultural, etc. Moreover, the development of geomatics should not be interpreted retrospectively as a coherent and distinct movement.

Since the 1990s, geographic information science has been established by striving to carve out its own research territory within existing disciplines (Maguire et al. 1991), not without violent reactions from the latter (Schuurman 2000). Professional geomatics has gradually structured itself as a specific and identifiable field through specialized companies, markets, budgets, experts, training structures and specific standards. While this movement is perceived as coherent from the inside, its structure is not as obvious from the outside. Its boundaries are blurred and shifting its core is difficult to define because it lies at the intersection of several disciplines: computer science, geography, information and communication sciences, sociology, etc. Its applications are potentially relevant to all fields. Furthermore, constantly evolving technology drives this general movement forward, leading us to talk in terms of evolution rather than revolution (Joliveau 2020). To drive the point home, some of the innovations that have most disrupted the field of geoinformatics come from outside the field itself. When Google launched its consumer mapping application in 2005, which greatly disrupted usage patterns, it was not a geomatics company. However, it enabled geomatics specialists to finally explain their mysterious work to an uninformed public in simple terms: "it's a bit like Google Maps, but more complicated". This "consumer" geomatics also broadened the spectrum of uses for geographic information. Within a decade, geovisualization became essential for mobilizing and organizing all types of content.

This evolution is well illustrated by one of Google's slogans used when it launched its mapping service: "Google Maps = Google in Maps". The inclusion of the word "in" is not anecdotal; it reflects the desire of the search engine invented by Larry Page and Sergey Brin to invest in geovisualization and turn it into an interface for accessing all types of information, provided that it is spatially indexed, or "geotagged" in web lingo. This moment marked the arrival of major IT and Internet companies in the field of geographic information, a trend that has continued to grow over time. These successive waves of technical innovation (databases, big data, the Web of Data, data visualization, artificial intelligence, etc.) are all moments that seem to herald the obsolescence of geographic information science. However, when the wave of novelty recedes, the specificity of spatial data reappears.

This introduction illustrates a curious paradox. Whether it is called geomatics in the field of industry and services, or GIS (Geographic Information Science/ Systems/Society) in research, this field remains highly specialized and relatively unknown. However, the tools involved are now widely available and are used for a great variety of purposes, often without our realizing it. Since the 1960s, we have witnessed a continuously accelerating process of transcribing into digital data all terrestrial realities, biotic and abiotic, all social and economic phenomena, all aspects of the organization and functioning of societies and the daily lives of individuals. A veritable "geodigitization" of the world is now underway (Joliveau 2007, 2020; Noucher 2023). The proliferation of maps on the web (Noucher 2017) is a visible manifestation of this, but it actually hides more profound transformations linked to the structuring role played by geospatial technologies[1] in society through

1 For simplicity's sake, in this book we use the terms "geospatial" data, information, or technology to refer to anything involving the digital processing of spatial or geographic information.

their management of location, distance and mobility. This accelerated digitization of all aspects of society coupled with unprecedented access to ever-increasing volumes of information, which is not only geographical, is giving rise to multiple concerns, debates and controversies. These include questions about security and control, surveillance and freedom, risk of addiction and loss of mastery, as well as systematic manipulation of opinions and attention. At a time when information technology has become so prevalent in society that a new word, "digital", must be used to describe this transformation, its geographical dimension has also become invisible. Yet geospatial technologies constitute the spatial infrastructure of new practices relating to space and movement and, more broadly, activities linked to information, knowledge and action on society and the world. However, the diversity of uses for these technologies, combined with their technical nature and specialized character, as well as their rapid and continuous evolution, makes them difficult for most users to perceive.

Unlike in other countries, geographic information science is not a separate field of teaching and research in France. While in the United States and Canada, for example, universities have geomatics departments (offering bachelor's, master's and PhD programs) and dedicated research centers, in France geomatics is taught in a variety of disciplines (geography, computer science, archaeology, architecture, ecology, etc.). This observation highlights both the strength and fragility of the field: its strength lies in being a truly new interdisciplinary field built at the crossroads of multiple research traditions, while its fragility comes from the fact that the research community needs constant networking efforts to flourish. This is the reason for the publication of books such as this one, which attempt to express the geographical, social and critical sensibility that characterizes this active but often isolated research community[2]. Faced with this paradox of ubiquitous but largely invisible geospatial technologies, we must not seek to confine this scientific and practical field to an illusory pre-established domain. We must both take stock of the current state of certain sectors of the field that have long been stabilized and open up new paths to lesser known areas. This is the objective of this book, which brings together authors of different generations. Some have been involved in geomatics or GIS research since its emergence, while others are in the middle of their careers or have just started out.

Chapter 1, written by Thierry Joliveau, outlines the establishment and general characteristics of the "geospatial factory", beginning with a review of the major stages in the deployment of geomatics and geographic information science and

2 The MAGIS research network (https://gdr-magis.cnrs.fr/) plays a key role in connecting French research in geographic information science. Several authors of this book are actively involved in this network.

systems. He describes the project underlying this enterprise, which aims to produce the most complete and accurate computer replication of the material and immaterial worlds and the way they work. He also points out the utopian and illusory dimensions of this project. This replication changes the way we perceive and understand the world. However, the geospatial factory also impacts individuals' lifestyles, spatial practices and social interactions, as well as physical environments. Several examples of this are provided.

In Chapter 2, Teriitutea Quesnot gets to the heart of the field by proposing a renewed conceptualization of geospatial data types that is more in line with the current conditions of their production on the web, where social networks now take the lion's share. To the classic distinction between "data" – data digitized from the physical environment – and "capta" – data captured by web platforms, often without their users' knowledge – he adds a third category, "quaesita", whose purpose is to use algorithms to influence individuals' spatial behavior. He then revisits the question of geospatial data through the concepts of geospatial traces and footprints and algorithmic spatiality. For him, the new techniques of augmented, diminished and mixed reality aim to superimpose these algorithmically calculated "quaesita" onto individuals' perceptions in situ to influence their decisions.

In Chapter 3, Boris Mericksay reviews OpenStreetMap (OSM), one of the most original and significant projects in the field of geomatics. In 15 years, it has grown from an amateur initiative to build an open world map to an internationally recognized source of geographic information used by a wide range of individuals, NGOs and public or private entities. OSM can be considered a digital "commons" unique in its history, nature, conceptual and technical choices, and its decentralized and open organization. What is perhaps less well known is the importance that OSM data have gained among professionals and operational staff through multiple partnerships with the public sector and active contributions from large private companies. Its unique and strategic nature makes OSM a key issue for the future of geospatial information.

In Chapter 4, Juliette Davret addresses the issue of the use of geospatial information in spatial planning, where methodological, theoretical and practical references are rapidly evolving. She takes the original perspective of maritime planning and its specific characteristics compared to land-based planning. Geographic information has become central to this field, and maritime planning uses it at all stages, with tension between a tendency toward standardization of the data used and an ever-increasing demand for adaptability to the local context. Geospatial technologies are also used to support the participatory dimension of planning processes. But the technical skills they require can lead to planning processes being reserved for experts in these methods. The uneven geodigitalization of the marine environment raises many questions that are also relevant to the terrestrial environment, where the homogeneity of cartographic coverage is only apparent.

In Chapter 5, Grégoire Feyt focuses on describing one of the oldest and most important uses of geomatics, namely, that which concerns local authorities and territories. He retraces the evolution of this "territorial geomatics", from the production of geographical data at the national level to the patchwork of data held by local authorities. He also questions its recent emergence as a "geocommons". In doing so, he highlights the disparities – both territorial and informational – that result from the transformation of public policy by geomatics. His analysis of the evolution of calls for tenders and changes in the job market suggests ways in which territorial sciences and geospatial technologies could be linked more proactively.

Throughout Chapter 6, Juliette Morel and Baptiste Hautdidier explore the more recent links between geomatics and digital humanities. Their contribution analyzes the convergence that has led to the emergence of the research field of digital spatial humanities. From the spatialization of digital humanities to the "humanization" of geographic information science, this overview, illustrated with research projects, raises a series of scientific barriers, both methodological and epistemological, which should structure part of the scientific community's work in the coming decade.

In Chapter 7, Allan Deneuville describes the role that geographic information plays in cyber intelligence and cyber security. Focusing on Open Source Intelligence (OSINT), the author identifies and describes several illustrative and analytical uses of geographic information. Based on journalistic investigations and NGO reports, his typology of uses allows him to reveal the different visual grammars at work and to highlight the mechanisms of visibility or invisibility associated with them. The political and conceptual blind spots he identifies also provide new perspectives for research.

Finally, in Chapter 8, Thierry Joliveau sets out to explore the sociotechnical imaginary of geospatial technologies. Based on an investigation of works of literary and cinematographic fiction, he attempts to approach the representations and images circulating in society about geodigitalization, which do not necessarily coincide with the ever-changing flow of technoscientific promises produced by the media. Approaching this imaginary through the prism of fiction is an original way of inviting readers to question the future of geospatial technologies.

This overview, far from being exhaustive, provides a broad picture of the uses of geographic information in various fields and by an increasingly diverse range of actors. The resulting research agenda is extensive. The conclusion emphasizes the need to break out of the technological somnambulism that characterizes our era and proposes developing and multiplying critical approaches in geomatics and geographic information science in order to better understand and transform the new "geospatial world" we are creating.

References

Joliveau, T. (2004). Géomatique et gestion environnementale du territoire. Recherches sur un usage géographique des SIG. HDR Thesis, Université de Rouen [Online]. Available at: https://hal.science/tel-04559188v1.

Joliveau, T. (2007). Géomatique et géonumérisation. Monde géonumérique [Online]. Available at: https://mondegeonumerique.wordpress.com/geomatique-et-cie/geomatique-et-geonumerisation/ [Accessed 6 February 2024].

Joliveau, T. (2020). Une révolution numérique de la géographie ? Le cas de la géomatique. *Histoire de la recherche contemporaine. La revue du Comité pour l'histoire du CNRS*, IX(1), 21–34. doi: 10.4000/hrc.4062.

Maguire, D.J., Goodchild, M.F., Rhind, D. (1991). *Geographical Information Systems: Principles and Applications*. Longman Scientific and Technical, Harlow.

Noucher, M. (2017). *Les Petites Cartes du web. Analyse critique des nouvelles fabriques cartographiques*. Éditions rue d'Ulm, Paris.

Noucher, M. (2023). *Blancs des cartes et boîtes noires algorithmiques*. CNRS Éditions, Paris.

Schuurman, N. (2000). Trouble in the heartland: GIS and its critics in the 1990s. *Progress in Human Geography*, 24(4), 569–590.

1

Understanding the *Geospatial Factory*

Thierry JOLIVEAU
UMR Environnement-Ville-Société, Université Jean Monnet de Saint-Etienne, France

1.1. Introduction

The issues addressed by geomatics and geographic information science are often perceived as essentially technical. They do indeed require a certain level of expertise in topography, mapping, computer science, data processing, statistics, etc. The manipulation of geospatial information follows precise methods and models and uses rigorous and specific protocols and formats. To practice it, it is necessary to master fairly complicated specialized software and have at least a basic knowledge of computer coding. However, this multifaceted and evolving technical dimension is only one element of a broader assembly designed to process geospatial information, which can be called the *geospatial factory*. It brings together men and women with a wide range of specializations from various organizations (public institutions, private companies, associations, informal groups, etc.). They have skills often acquired in universities and schools, which renew them through research. All their activities are regulated by bodies that produce laws, rules, standards, etc. This factory has been gradually built up over more than 50 years to carry out an almost continuous process of digitalizing geographic information, which we will call *geodigitalization*. It is heterogeneous, riddled with conflicts of interest and disagreements over methods and priorities. It has no clear boundaries. It has changed significantly over time in response to technical innovations and societal transformations, and along with the

Geospatial Technologies and Society,
coordinated by Thierry JOLIVEAU and Matthieu NOUCHER. © ISTE Ltd 2026.

increasing scope of geodigitalization. It would take several books to present this digitalization process of geographic information in any depth. But it is possible, as we will attempt to do in this chapter, to reconstruct retrospectively what constitutes, behind a multitude of more or less coordinated initiatives, the core of its project: the systematic digital transcription of geographic information covering the whole world. The presentation is in three sections. In section 1.2, we restore a sense of history to the geodigitalization process by isolating a few major milestones that marked the deployment of geomatics and geographic information science and systems. Section 1.3 focuses on the powerful but little-discussed driving force behind this process: the temptation to replicate the world through computer simulation. Finally, in section 1.4, we seek to show that geodigitalization is not just a matter of representing the world (material and unmaterial). It must be seen as an operating matrix that not only contributes to changing perceptions of reality but also interferes with the spatial structure of social interactions and behaviors, and materializes in the physical environment. Examples from three areas illustrate these effects: the role of digital geolocation data in sports, mixed reality gaming and management of Covid-19 pandemic; the use of big data in transportation; and the geospatial logic of social interactions at the hyperlocal level.

1.2. The making of the *geospatial factory*

The steps below are milestones in a complex story that correspond to key moments in the geodigitalization process.

1.2.1. *GIS*

Until the late 1980s, digital geographic information technologies were confined to university laboratories, mainly (but not exclusively) in the United States, where researchers developed geographic information systems (GIS) (Foresman 1998; Chrisman 2006). These research tools were then transformed into standalone software sold by private American companies, which quickly conquered the market of businesses, local authorities, administrations and research centers. Their rapid adoption by researchers led to a heated debate in English-language geography at the time, which contributed significantly to shaping the future field of research in geographic information science (Schuurman 2000). In France, these new tools have also triggered hostile reactions from geographers on two historically opposed sides (Joliveau 2004). On the one hand, traditional geographers criticize the excessive importance given to technologies that they consider to be nothing more than geographical engineering, distracting users from true geography, which is conceptual and discursive. On the other hand, proponents of a new theoretical and quantitative geography already use computers and have often developed their own programs to implement formal modeling and a nomothetic approach. The latter are also highly critical of this new software, which they see as tools without theory, ignoring the foundations and recognized models of spatial analysis and the established semiological principles of cartography, and offering only simplistic geometric combination operators with an essentially idiographic purpose (Rimbert 1989; Groupe Dupont 1994). Both sides agree that these trendy "GIS software" programs are suitable for operational issues but not for fundamental conceptualization. The operational dimension of these computer technologies contributed to their success in the 1990s in what has been known in France since the 1960s as "applied geography" (Philipponneau 1960). Between 1960 and 1980, and even beyond, this form of geography, which was geared toward solving practical issues, led many geographers to pursue careers in town and country planning. Therefore, it was in the departments responsible for land management and environmental planning that GIS quickly became established as a means of automating a number of tasks, mainly cartographic ones. GIS were also quickly adopted in ecology, hydrology and environmental sciences as a complement to specialized digital tools.

1.2.2. *Geoinformatics*

At the time, it was not widely understood that this wave of computerization was not limited to the adoption of software for automating certain repetitive and tedious mapping tasks and visualizing the new images captured by the first Earth observation satellites. All these data, painstakingly entered by hand on A0-size tables or retrieved from magnetic tapes, formed a system, as the name GIS indicated (Joliveau 1996). It constitutes a systematic, organized and generalized transcription in the form of information on facts, observations and knowledge concerning all phenomena detectable on the Earth's sphere.

This ongoing computerization has led to the creation of large databases for analysis. Contrary to George's (1978) assessment at the time, these data have increasingly being used to support decision-making in all fields. GIS represent only the geographical aspect of this general process of recording and using computer data. Certain fields of primary interest to geography have remained generally wary or even hostile to this computerization process, such as urban planning and landscape management, where human expertise in contact with the field and stakeholders has remained the benchmark for information gathering.

1.2.3. *Geomatics*

What has been called geomatics or geographic information science since the 1990s actually refers to the gradual reorganization, through the concepts and methods of computer science, of the knowledge and techniques related to the acquisition, storage, processing, analysis and dissemination of spatially referenced data. Previously, these tasks fell under related but distinct disciplines, such as topography, cartography, geodesy, photogrammetry and remote sensing. These disciplines did not merge because they correspond to different professions, skills and needs. However, as geographic information becomes more digitalized, it becomes easier to connect the various activities and applications related to this information. Geomatics links together all the technical systems related to geographic information. These systems are becoming increasingly numerous and diverse as the concepts and methods of GIS and automated cartography begin to circulate more widely among experts interested in their use in an increasingly varied range of fields: archaeology, ecology, transportation, the environment, urban management and planning, archaeology, etc. (Joliveau 2007).

1.2.4. *Geoweb and neogeography*

The emergence of the Internet in the mid-1990s completely transformed digital forms of access to geographic information (Scharl and Tochtermann 2007). The creation of online map portals reversed the traditional relationship between data and maps. Databases are no longer queried to produce maps; instead, cartographic interfaces are used to navigate an increasingly rich universe of data, leading to the harmonization of formats and the structuring of forms of cooperation between data producers and holders (Joliveau 2011). This networking of databases, applications and users gives the general public access to an enormous volume of geographic information: base maps, satellite imagery, thematic data, etc. It is transforming the processes of map production, distribution and use (Mericskay 2016). Spatial indexing of all types of digital content is becoming widespread to organize access and enable data retrieval. The techniques used are georeferencing, geocoding and geographic indexing. Mapping methods on the Internet are also changing, increasing the ability to navigate at different scales and select and aggregate information (Guérois and Madelin 2024). A whole new world of interactive, dynamic and animated geovisualizations is emerging (Cunty et al. 2017). The so-called Geoweb is neither unified from a technical nor from a usage standpoint. The technical process of acquiring, storing and consulting geographic data is made up of numerous protocols. And geographic information has become a significant market, with multiple services seeking a competitive advantage, or even a monopolistic position.

It is also the moment when new global players, such as Microsoft and, above all, Google, enter this market.

In the mid-2000s, computer technologies evolved, and the Internet transformed from a medium for finding information to a medium for self-expression, through comments on websites, contributions in forums and blogs creation. Personal mapping applications accessible on the go are emerging, initially with Wi-Fi-enabled laptops, then with connected smartphones. The web became cartographic following the success of Google Maps/Earth, which allowed users to view search engine results, navigate freely on a single world map and create personal maps. OpenStreetMap[1], created by Internet users, can be considered a free version of Google's proprietary map or the state maps produced by national mapping agencies. This contributory shift in geographic information is changing the public's relationship with geographic information. It was quickly proclaimed a major revolution called "neogeography" by some of its players (Turner 2006). In fact, it turned out to be a redistribution of the economic and industrial system around the production and exploitation of geographic information, which placed large web companies, particularly Google, at the center of the field (Joliveau 2010). This cartographic effervescence resulted in the emergence of a new form of mapping that is more fragmented and scattered (Noucher 2017). In research, this contributory dimension of geographic information was theorized very early by Goodchild under the name VGI (Volunteered Geographical Information), which invites us to understand how individuals and citizens participate in the creation of geographic information (Goodchild 2007).

1.2.5. *Social (geo)media*

This desire among Internet users to contribute was quickly picked up by the social media platforms that were emerging at the time. Facebook (created in 2004) and Twitter (created in 2006) were the leading examples, but many others have followed since: Instagram, Snapchat, TikTok, etc. These platforms have become the preferred channels of communication for billions of people around the world, at the expense of forums and blogs, which have lost their appeal. They constitute information spaces of unprecedented size and intensity, which have become genuine geopolitical objects. Under the guise of a horizontal connection tool, social networks organize and filter access to information through algorithms. These are geared toward capturing and maintaining the user's attention, imposed by an economic model based on free service in exchange for the provision of personal data that feeds

1 See B. Mericskay's analysis of OSM as an example of digital commons in Chapter 3.

personalized profiles sold to advertisers. The location of users is quickly perceived as an important feature of these profiles by social media applications. At the end of the 2000s, social networks were created specifically around sharing geographical location: Google Latitude, Foursquare and Gowalla, among others. This led to predictions of the emergence of a new type of "geomedia", which would allow individuals to continuously share their experience of immediate environments and places visited using geolocated and multimedia information such as texts, images and sounds[2]. In fact, these new social networks do not last. Dominant social networking platforms, such as Facebook (which bought Gowalla) and Snapchat, incorporate the social sharing function of geolocated information. Foursquare continues to develop geolocation-related technologies but in 2014 transformed itself into an urban guide for peer-to-peer recommendations of interesting places. As Nova (2009) notes, the emergence of autonomous geomedia is short-lived. However, this episode leads researchers to wonder whether GIS should be theorized as media as much as information systems (Sui and Goodchild 2003; Crumplin 2007). The interpenetration of geospatial technologies and social media also invite us to cross-fertilize geography, which is opening up to communication issues, with communication sciences, which are negotiating a spatial turn. Sui and Goodchild point out a number of concepts that geographic information science should re-examine in light of this encounter with social networks. This project is still ongoing: it includes visualizing interpersonal networks, exploring the multimedia dimension of data and better conceptualization of what a place is in GIS models initially built to account for space (Sui and Goodchild 2011).

Several other milestones could be added to this short history of the geospatial factory including satellite imagery, big data, the Internet of Things and, more recently, artificial intelligence. The proposed markers clearly outline the stages of the digital transcription process of geographic information.

1.3. Geodigitalization as replication

Since the mid-1960s, Geomatics and geographic information science have therefore been at the heart of the geodigitalization. This process employs various computerized methods and direct or indirect georeferencing techniques to systematically transcribe all objects, beings, phenomena, activities, artifacts and creative works. This process, which seems essentially technical, has taken on different forms over time. But it has mobilized and continues to mobilize dreams and fantasies, utopian goals and fundamental concerns, which are worth revisiting.

2 Quesnot tackles this issue head-on in Chapter 2.

1.3.1. *Replicating the physical world*

Geodigitalization initially involved transferring existing maps, aerial photographs and satellite images of concrete objects of the physical world – roads, rivers, houses, terrain, etc. – to computer media with ever-increasing precision and detail. Intangible objects such as property boundaries, which are particularly important in Western societies, were also digitally transcribed from the outset. These concrete objects have been associated with characteristics related to their nature (type of road, building, etc.), their use (number of vehicles passing on a road, etc.) and their functioning (flow rate of a watercourse, etc.). As it expanded and became more precise, this digital description came to encompass industrial devices and entire urban organizations. This georeferenced description of the world is constantly expanding and becoming increasingly detailed and up-to-date. It includes natural atmospheric, geophysical, terrestrial and oceanic phenomena, natural environments, and surveys of living organisms, plants and animals, as well as measurements of demographic, economic and social phenomena. The geomatics project has gradually emerged as the most faithful reproduction possible of the physical world – a replica of it, but always in the form of explicit, formalized, documented and shared models.

Admittedly, the multiplicity of sources, technical devices and operators means that this digital double is not an exact, seamless copy of its physical model. The replica is imperfect, cracked, irregular and uneven. But since its inception, geomatics has strived to coordinate initiatives and promote the interoperability of formats and tools to give coherence and continuity to this digital copy of the world. This is despite ongoing competition between stakeholders and differences linked to the monopolistic tendencies of the large companies involved, as well as the fierce determination of institutions to protect their turf.

1.3.2. *Georeferencing unmaterial and mental objects*

Hill (2006) was one of the first to point out that the ongoing systematic georeferencing was extending beyond the material world to include immaterial entities such as cultural and artistic works, media content, images and texts, and everything related to heritage[3]. Geodigitalization now encompasses all human communications, fundamentally changing the status of geospatial technologies. Until then, these technologies were a separate domain, a well-defined field of expertise derived from highly specialized professions and disciplines responsible for

3 Chapter 6 of Hautdidier and Morel clearly illustrates the conceptual and theoretical renewal produced by the convergence of geomatics and digital humanities.

providing a robust and documented description of the state of the world. In digital mode, they have become part of the most intimate and everyday aspects of life through social media and ambient computing technologies with their ever-increasing number of sensors and connected objects.

1.3.3. *When cyberspace reverses itself*

A major shift in this geospatial world can be seen at the turn of the 2000s, when computer technology had already advanced enough to raise the question of its visual representation. This was the moment when visualization techniques, driven for 20 years by the rapid development of video games, became mature enough to render this mass of computer data in three dimensions. The renderings were sufficiently convincing and realistic, even on standard computers, to raise fears or hopes that this digital world would compete with the physical world. The visual similarity between the replica and its model coincided with the release of realistic open-world video games such as *Grand Theft Auto*, which digitally recreate[4] real locations. These were also the years of the launch of Second Life, the first persistent virtual universe where participants could interact through their avatars. It may come as a surprise to learn that Second Life still exists in 2025, while Meta's ambitious attempt to create the Metaverse ended in spectacular financial failure in 2023. However, the emergence of this gesospatial world, topographically aligned with the physical world, seems to be, in the words of novelist William Gibson[5], as if cyberspace were no longer "a specific elsewhere, one we visited periodically, peering into it from the familiar physical world. Now cyberspace has everted. Turned itself inside out. Colonized the physical"[6]. Several researchers have attempted to conceptualize this reversal, which directly links physical locations to resources on the web and, conversely, web "sites" to real locations. Zook and Graham (2007) proposed calling *Digiplace* this hybrid place, both digital and physical, which we inhabit or visit today. The local, physical environment of a place is complemented by a fluid, constantly updated "cloud" of information that refers to and describes it. We can explore a place physically and/or digitally and use georeferencing to match the cloud of information with the physical space. Digiplace is always constructed by a computer program that queries the data available on a place. Visiting the *Digiplace* is interactive and individual, as it depends on the code/program used, the user's skills and their personal interest. Not all digital spaces are equal, either in terms of their physical or social characteristics

4 By significantly altering the topography of the area (Joliveau 2012).

5 Creator in 1982 of the name and first imaginary description of cyberspace in his novel *Neuromancer*.

6 http://www.nytimes.com/2010/09/01/opinion/01gibson.html. See Chapter 8.

or in terms of the accuracy, completeness and detail of their translation into data. It is important to note that this cyberspace, which has become ambient, is not virtual in the common sense of being unreal. According to Vial (2014), inspired by Turkle (1995), it is a "computer-simulated" space, a digital space that is added to and combined with physical space. As Vial writes, geodigitalization is gradually leading us to change our perceptual structures. We now are present to things and beings also as digital interfaces show them to us (Vial 2014).

1.3.4. *Digital twins of territory*

In recent years, the creation of a geospatial replica, a "computer-simulated" version of the world, has continued to develop in a new technopolitical context: the digital twin. This concept was imported from industry in the context of the smart city. The aim is to digitally reproduce an object, process or system as accurately as possible to simulate how it functions. After 3D models of a territory, which focused on replicating the visible dimension, the goal is now to create digital replicas of the territory "as it works". A digital twin of a territory can be defined as:

> A dynamic and intelligent digital model of the city that integrates static and dynamic (real-time) data, both public and private, into a single reference environment. It is a consolidated information system that integrates business tools (mapping, planning, e-administration, etc.). The 3D visualization offered by the digital twin facilitates the contextualization and understanding of urban data. (Banque des Territoires 2021)

The digital twin theory is very much a product of the smart city. It shares some of the same beliefs and assumptions, which are sometimes paradoxical. The digital twin is based on a functionalist approach that treats the territory – an anthropic, organizational, social and political entity – as a complex technical system. It considers that centralizing data will resolve the separate siloed approaches that are believed to be the main source of territorial dysfunction. It presupposes that intensifying real-time data collection and artificial intelligence technologies, a kind of local geo-engineering, will make cities more sustainable and frugal. The concept of digital twins is not limited to the city scale. The European Commission's "Destination Earth" project, for example, aims to develop a digital twin of the Earth "to model, monitor, and simulate natural phenomena, risks, and related human activities".[7]

7 See: https://destination-earth.eu/.

The digital twin appears to be the ultimate utopia of successful geodigitalization. Not only will all the necessary data needed to model all processes be collected, but it will be possible to model and simulate all these processes and their systemic relationships by integrating complexity sciences (Arcaute et al. 2021).

Lucas (2022) clearly highlights the imaginary dimension of the digital twin of the territory through the many myths it revives: Pygmalion, Narcissus, Prometheus, Faust, Babel, Sisyphus, Gaia, etc. Apart from the dizzying dream of merging the object and its representation, there is still the risk of granting more power to experts and engineers. They are the only ones able to understand the models that lead to decisions.

Furthermore, our social interactions are already driven by algorithms and our itineraries calculated by navigation systems. We know in real-time whether a place is busy and what time it will stop raining. An artificial intelligence tool can now write a novel set in the places we love or draw maps of the landscapes we dream of living in. The ambient and ubiquitous "digital twin" that would orchestrate this utopia takes on a "Big Brother" air.

1.4. Geodigitalization as an operating matrix

The geodigitalization of the world cannot be assessed solely in terms of data and representation. The widespread adoption of geodigital technologies and data is transforming the way we represent the world – but its impact is broader than that. In subtle yet fundamental ways, it is changing how we explore, understand, experience, talk about and critically engage with the world to transform it. Geospatial technologies are now an integral part of contemporary social life, having assimilated and even reinforced existing forms and structures. They have also created a new way for people to access the world, leading to new behaviors and uses as well as new opportunities and risks. We must consider the implementation of geospatial technologies as *operations* to use the term employed by Besse and Tiberghien (2017) in relation to cartography. These operations transform the ways we do, act, and think.

The process of geodigitalization is widespread. Its effects are subtle but affect virtually all domains. Here are a few examples of these geospatial operations and the transformations they facilitate and accompany. Some are global and massive, others seem more modest and innocuous, but all appear to be irreversible. These are just a few highlights; each issue would need to be addressed in greater depth.

1.4.1. *Digital geolocation, from entertainment to control*

Real-time geolocation functions have become the building block of many applications based on what Beaude (2012) calls "synchorization", that is, the synchronized spatialization of actions. These applications are used to connect people most often by turning them into users of a private digital service that acts as an intermediary. The implementation of these synchorization services has greatly transformed many aspects of daily life in wealthy countries and elsewhere: travel route planning (Google Maps, TomTom, Waze), individual transportation organization (Uber), tourist accommodation rental (Airbnb), dating (Grindr, Tinder), access to healthcare (Doctolib) and so on. The economic, social and spatial effects of these highly popular services have already been widely documented. What is less well known, or often omitted, is that the geolocation data collected is used for other purposes. We have deliberately chosen two examples of use that relate to very different situations, but both have global implications: sporting and leisure activities on the one hand, and the Covid-19 pandemic on the other hand. The contrast between individual activities and centralized coordination at the national level illustrates the extent to which digital geolocation technologies have become ingrained in our lives.

1.4.1.1. *Geolocation of sports and recreational activities*

A first example of a recreational application is Strava, a social network linked to GPS, widely used by runners and cyclists who record and share their journeys online to compare their performance with other users. It became very famous beyond the sports world in 2018 when an Australian student revealed that soldiers' geolocated Strava tracks were freely accessible online in the form of a heat map. This allowed other users to locate US military bases in Syria and Afghanistan. In densely populated areas, individual anonymized tracks were aggregated into the most frequently used routes, but in desert or less frequented areas, they were clearly visible. It then emerged that the Strava website gave access to users' public profiles, allowing them to be identified, and their home locations to be deduced from their tracks. In 2024, the French newspaper Le Monde revealed that many bodyguards of heads of state continued to publish their running routes on Strava with a public profile under their real names, making it easy to locate where these world leaders were staying (Untersinger 2024). Strava pointed out that users could limit the visibility of their profile and activity to their friends, or even make them private. However, subsequent investigations showed that these "sensitive" users had neglected to do so.

Mixed or alternate reality games are another recreational use of geolocation, made famous by the global success of *Pokemon Go* in 2016, where players capture digital Pokemon by exploring their surroundings using their smartphone screen. The game was developed by Niantic, a Google subsidiary that became independent in 2015. It is based on the same computational and spatial architecture as *Ingress*, Niantic's previous game. In Ingress, players worldwide are divided into two teams and compete for "Portals" via their smartphone's digital interface. These portals are associated with physical locations, many of which were created by players. Launched in 2012, the game was integrated into Google's geospatial ecosystem alongside Google Maps, Google Earth and Google Street View. It enabled the company to collect on-site data about places and geographical features that are meaningful and emotionally significant to users. *Pokemon Go* took over *Ingress's* geographic database, with Portals becoming PokeStops. In March 2025, Niantic sold *Pokemon Go* to Scopely, a smartphone video game company owned by a Saudi investment fund. Niantic's artificial intelligence division became Niantic Spatial, an independent company in which Scopely invested. Niantic Spatial used *Pokémon Go* data to train a "large geospatial model (LGM)" based on large language models (LLM) on textual data, which is intended to "help computers perceive, understand, and navigate the physical world" (Brachmann and Prisacariu 2024). Players had not anticipated this use of their geolocation data: contributing to the creation of a private company's database, which the company then uses to train artificial intelligence. It is also an original illustration of the economic value of geographic information through gaming (Odobašić et al. 2013). Incidentally, given the problems of LLMs trained on biased data, we can anticipate issues with this future LGM. Colley et al. (2017), for example, found that in 2016, Pokestops were significantly more numerous in urban, white, affluent neighborhoods in the United States.

Geolocation data is collected by applications for which knowledge of a geographical position is essential, such as dating apps (Grindr), peer-to-peer goods sales apps (Leboncoin, Vinted), and weather forecast apps (Météo et Radar). Many other personal apps that are indifferent to location also record this data, opening up privacy and security breaches similar to those seen with Strava. This geodata can also be purchased through data brokers, companies that control this highly opaque market (Sénécat et al. 2025). Even more worrying, a tool provided by the company Babel Street is indicative of the advent of a society of pervasive control and surveillance. This tool called *Locate X* uses a polygon-drawing function on a map of any location in the world to list and visualize the advertising ID of the mobile phones that have entered a specified area in the previous days. It is then possible to identify the home address of these individuals by searching for the night-time locations of their phones or by directly purchasing files that link the mobile advertising ID to the name, address, email and phone number of its owner (Kebs 2024). The legality of this type of software tool is being debated in the United States, but there are growing concerns that it could be used in an uncontrolled manner by private organizations or police forces for individual surveillance purposes.

1.4.1.2. *Geolocation and Covid-19*

The Covid-19 pandemic in 2020 was a global phenomenon that led several countries to discuss the advisability of using individual geolocation data to control the epidemic, through tracking the contacts of infected individuals and monitoring people's movements (Boudreau and Caron 2022). There were significant differences between countries in their adoption of this type of policy. In South Korea, the authorities very quickly set up a technological platform called the Epidemic Investigation Support System (EISS) adapted for epidemiological purposes from a pre-existing system developed as part of a smart city project. This platform centralized individual data from multiple sources: credit card transactions, cell phone geolocation, health records, public transportation travel, surveillance cameras in public spaces, etc. Its purpose was to identify and notify people who had been in contact with an infected person and who had not worn a mask or respected physical distancing and to request that they self-isolate. The platform cross-referenced, analyzed and mapped all these data in a very short time. France, Canada and Quebec opted to develop applications for use on a voluntary basis. These applications were designed to anonymously track Bluetooth connections with other users, so that anyone who had come into contact with someone who had reported the disease could be notified. These apps were based on the principles of non-centralization, non-identification and non-localization. To function effectively, they needed to be adopted by a significant portion of the population. In none of these three countries did these applications play the expected role. Determining the relative effectiveness of the two systems is not easy. A thorough investigation into the reasons why a country has or has not opted to track individuals during the pandemic would be necessary (Guillaud 2021). A number of factors may explain the choice between these two approaches such as the more or less individualistic nature of the societies in question, the degree of digitalization of individual practices, the level of deployment of ubiquitous computing and smart city projects, the nature and age of the relationship with democracy. For our purposes, this is a real-life demonstration that geospatial technologies are already being widely used and are not merely limited to a simple collection of geolocation data. A comprehensive system of spatial data information and analysis can be mobilized for the purposes of state surveillance and control of individuals. The absence of a systematic and uniform geospatial control framework in the context of the Covid pandemic highlight the importance of the political and cultural dimension in this domain. It underscores the challenges associated with democratic oversight of the use of these technologies, which authoritarian states may exploit for purposes that undermine freedom. Another risk stems from the fact that digital technology companies have become as powerful as, if not more powerful than, governments. They are opposed to the regulations that states are trying to impose. At the same time, the market for personal geographic data, whether legally obtained or not, is growing very rapidly without real control.

1.4.2. *Material effects of geolocation in transportation*

The ability to remotely and in real-time geolocate individuals and objects can have significant impact on individual behavior and social relationships. It also influences concrete actions and infrastructures in various ways, as demonstrated by three examples from the transportation sector.

1.4.2.1. *Planning*

The first example describes a purely potential effect, when geolocation data are used to plan physical infrastructure. With more than 135 million claimed users in 190 countries, the geolocation data available on the Strava app mentioned above provides a basis for analysis by researchers and planners. These data can be used, for example, to better understand the practices of cyclists in cities and plan appropriate amenities. Battiston et al. (2023) examine the reasons why fewer women than men cycle and attempt to understand whether this could be due to physical conditions and cycling facilities. Using Strava data, they calculate a ratio of female to male cyclists and statistically compare it with indicators describing the local physical environment. The local physical environment is defined by factors such as road network structure, types of intersections, speed limits, presence and types of bike lanes and air quality. This comparison is conducted at the city and street levels in various American and European cities. The study's findings highlight discrepancies among countries and validate most of the hypotheses concerning safety, opening up a new interrogation on women's perception of safety in cycle lanes that are not physically separated from traffic. However, the authors conclude that the main limitation of their study is the non-representative nature of Strava data. Strava users tend to be sports and leisure cyclists, whereas the facilities are primarily intended for people who cycle to school and work on a daily basis. The bias in geolocation data from big data may therefore limit its use in planning.

1.4.2.2. *Infrastructures*

The second example in the field of transportation concerns the effective impact of geolocation on urban planning through the use of geofencing techniques for free-floating bicycle and scooter sharing platforms in the Île-de-France region, as highlighted by Mangeart (2023). Free-floating vehicles do not have a station where they can be picked up and dropped off. Geolocation is therefore essential to enable customers to easily find a vehicle in public spaces and for the company to maintain them. Both customers and technicians use smartphone apps to manage the vehicles. In Paris, several platforms have been offering this type of service since 2016. They initially took advantage of the regulatory uncertainties at the time to launch their services without seeking authorization from local authorities. The presence of these new vehicles quickly became controversial and led the city of Paris to introduce "codes of conduct". In 2019, the "Loi d'orientation des mobilités" (Mobility Orientation Act) provided the regulatory framework for negotiating two major issues: the occupation of street space and the sharing of digital data collected by platforms.

In fact, these two issues go hand in hand, as a significant portion of the data is geospatial and describes the space traveled by vehicles, which is the central focus of the negotiations. Geofencing functions have been the means by which scooter platforms have developed a strategy for managing this space. Geofencing consists of drawing on a digital map in geographic information software the areas where, by computer programming, scooters must stop, slow down or be dropped off. For example, the City of Paris wanted to reduce the maximum speed of scooters from 20 to 10 km/h in certain areas of the city but was unable to impose this on the platforms. However, the platforms did eventually agree to the gradual introduction of dedicated parking spaces for scooters, outside of which users could not end their journey and leave the scooter. The platforms resisted the introduction of a feature limiting the maximum number of parked scooters, fearing that it would discourage users who might not be able to return their vehicles to their desired location. The physical space of the street is therefore subject to negotiation through its digital transcription, which becomes directly performative at the level of the scooter itself. In this situation, the community's prerogative over physical space is balanced against the power that geospatial data, which only the platforms possess, gives them. However, as Mangeart (2023) points out, the forces are not balanced, especially in small communities. Geospatial skills and tools are on the side of the platforms, which spread their investment across several cities, while communities do not have the means to equip themselves with geofencing tools. It is therefore the platforms that communicate to local authorities the boundaries of the parking areas entered into their system. They also have data on the spatial flows of scooters and on rule violations. They refuse to share this data, citing reasons of cost and underuse by local authorities. However, confidentiality with regard to competitors also plays a role. In practice, free-floating vehicle platforms are in a situation of quasi-self-regulation: "Thanks to their mastery of digital tools, they can take liberties with regulations that local authorities do not have the means to enforce" (Mangeart 2023). This example of Parisian scooters may seem isolated and anecdotal, however, the report by Hansen et al. (2021) shows that the implementation of geofencing measures is spreading across Europe. We can see how digital geolocation translates into both a physical parking space and direct digital control of behavior. Without prejudging future developments, these solutions could become widespread with the growth of shared or autonomous vehicle fleets in the context of traffic regulation in high-risk situations (congestion, pollution, flooding, etc.) or in other situations where access is restricted.

1.4.2.3. *Simulation*

The third example concerns the Waze navigation system, which results in hundreds of cars driving through quiet residential areas in search of the fastest route. Waze is a GPS-based smartphone navigation system developed by the Israeli company Mobile and acquired by Google (Alphabet) in 2013. It claims to have 140 million users worldwide, including 20 million in France. Its unique feature is that it relies on a road network provided by users who can report events or problems during their journeys, which are then received by all connected users in the vicinity. The route is recalculated instantly and very quickly based on traffic conditions. As a result, in the event of traffic jams or roadworks on major roads, Waze tends to redirect traffic to smaller roads or quiet residential streets nearby. For years, groups of residents have been mobilizing against Waze in various ways. The first was digital, reporting on Waze false traffic jams or non-existent incidents on the street concerned. This did not work because more drivers confirmed the new route, and Waze ended up blocking residents' accounts for bad behavior (Furno 2016). The reactions took a physical turn with anti-Waze signs at the entrance to villages. Then the municipalities introduced speed limits, which could work, not necessarily because drivers respected the limit, but because once the maximum speed for the segment was integrated into Waze, the algorithm no longer suggested the route. A one-way system or speed bumps on the road could also work. These are all physical responses to a geospatial problem.

Courmont (2018) explains the nature of the problem very clearly. Waze calculates road traffic based on geolocation and real-time reports from millions of drivers connected to the app across the entire network, from highways to small streets. Traditional methods used by road network managers measure traffic flows at fixed points on their network. These two models differ in their logic but also in their objective. Road managers regulate traffic flows across their hierarchical network so that, collectively, overall traffic flows as smoothly as possible. Waze aims to optimize the individual travel time of each connected driver. In addition, unlike road network managers, it has real-time information on the entire network traveled by connected cars. In the event of a problem on the main road network, it distributes traffic to the secondary network, including residential roads. These two models are actually complementary, as the new route suggested by Waze may not ultimately result in the time savings expected by the driver, given changes in overall traffic flow. Furthermore, Waze does not have information about roadworks or events that block or slow down traffic before its users discover them. In fact, with its Connected Citizens program, Waze offers local authorities access to its traffic data, obtained from the geolocation of its users, in exchange for information about roadworks and street closures. As Courmont (2018) points out, this exchange of data is also an exchange of legitimacy. Local authorities strengthen their ability to regulate traffic by integrating data that they cannot produce themselves. Waze strengthens the trust of its users and becomes part of the official public decision-making system. This example also illustrates how geolocation data always refers to a state of the physical environment that it gradually helps to transform if the data can be aggregated.

1.4.3. *The geospatial logic of local information*

What role do geospatial technologies play in the way individuals manage their local information online, with regard to their living space and place of residence? Numerous tools handle this abundant and highly diverse information. The first local websites to find a wide audience were those offering classified ads between individuals, such as Craigslist in the United States, created in 1996, or Leboncoin in France. To share more regular and diverse local information online, many digital tools also offer solutions: social networks (Facebook, Instagram, Twitter/X), messaging services (WhatsApp), forums (Reddit) and networks specializing in neighborhood connections (Nextdoor, AlloVoisins, etc.). This technological proliferation demonstrates the importance of information exchange on a local geographic basis. It also highlights the difficulty of understanding its scope and nature, given its dual compartmentalization by territory and platform.

1.4.3.1. *Facebook groups and Nextdoor neighborhoods*

Two examples are particularly significant: Meta's Facebook with its local groups, and Nextdoor, the hyperlocal social network run by the company of the same name. In 2010, Facebook gave its members the option to form groups based on shared interests, which led to the creation of local, geographically based groups. There is a total of 10 million Facebook groups worldwide, but there are no figures or definitions for specifically local groups. Herdağdelen and Adamic (2023) were able to use Facebook's internal data to show that the distribution of Facebook groups (local or otherwise) differed regionally in the United States in relation to major sociocultural parameters. According to Zahnow et al. (2024), local Facebook groups in the Brisbane region of Australia are also more numerous and active in affluent, ethnically homogeneous neighborhoods and along the more touristy coastline. At around the same time as local Facebook groups, Nextdoor was created, the first and largest hyperlocal social network focused on neighborhood connections. In 2025, it claimed to have 340,000 neighborhoods and more than 100 million neighbors in 11 countries, including France[8]. Thirteen percent of Americans were reported to be using Nextdoor in 2021 (Auxier and Anderson 2021). According to Iqbal et al. (2023) and Brown et al. (2024), American Nextdoor neighborhoods also tend to be areas where the population is wealthier than average, whiter, more educated and more likely to own their homes. Local Facebook groups and Nextdoor neighborhoods are therefore very similar. They are both owned by large American digital technology companies and aim to meet broadly the same local needs: information sharing, commercial and non-commercial exchanges of goods and services, mutual aid and assistance, neighborhood watch, etc. This similarity likely explains Meta's decision to create Neighborhoods in 2021 in the United States and Canada, a specific feature dedicated to neighborhood relations similar to Nextdoor[9].

1.4.3.2. *Two spatial logics*

However, the two companies operate differently. Creating a local Facebook group is free and its geographical scope is not limited. The group is located using a name freely chosen by its creator, and membership is voluntary. Nextdoor, on the other hand, is based on a predetermined network of areas. The company sets the names and boundaries of neighborhoods, which are displayed on a map. Members are assigned to a neighborhood based on their address, but they also have access to information about adjacent neighborhoods. As a result, while several Facebook groups may coexist in the same space, all services offered on Nextdoor are centralized in the user's Nextdoor neighborhood, including the creation of specific interest subgroups similar to those on Facebook. This difference in organization may explain why the US counties dominated by highly local Facebook groups, studied by Herdağdelen and Adamic (2023), are characterized by a strong presence of local organizations: schools, libraries, scout groups, churches, volunteers, etc., each of which likely has its own Facebook group. In his survey of the Ain department, Trably (2024) confirms for France this link between local Facebook groups and associations or organizations that operate in the area and are the main sources of consultation and dissemination of local information.

In contrast to this plurality of perspectives on an undefined space, the Nextdoor model takes a more territorial approach, as illustrated by the example of San Francisco in 2016 (Payne 2017), when users were able to determine the boundaries of their own neighborhood. Amidst large neighborhoods with historical

8 See: https://about.nextdoor.com/.

9 Meta has said little since then about this feature and how it has been received. It does not appear to have been rolled out to other countries.

or toponymic names (Tenderloin, Lafayette Park, etc.), two tiny "neighborhoods" were added, each corresponding to a luxury apartment building. It was indicative of a small group of wealthy people who aspired to maintain a sense of exclusivity and distinction from their surroundings. This territorial demarcation may also explain the significant controversy surrounding Nextdoor and the prominence of issues such as insecurity, crime and racially based reporting of suspicious activity in the United States and Canada (Kennedy and Coelho 2022; Nextdoor 2024). These issues have led to Nextdoor neighborhoods being described as "digital gated communities" (Kurwa 2019). The reality seems more nuanced, as shown by Brown et al. (2024) for Nextdoor neighborhoods in the United States and Zahnow et al. (2024) for Australian Facebook Local Groups dedicated specifically to neighborhood watch. In addition to the logic of segregation, the minor yet systematic role of surveillance in these digital neighborhood exchanges, and above all the visual and direct nature of its exercise, once again reveal the practical, material and concrete nature of the information shared in these hyperlocal communal spaces[10].

1.4.3.3. *Imprecise geolocation*

There is no need for precise real-time geolocation on this type of geosocial network. On Facebook, geotagged posts or photos mainly concern unfamiliar vacation destinations and very rarely local areas (Trably 2024). On Nextdoor, messages of all kinds are exchanged and filled with topographical indications, but these often remain somewhat vague. Neighborhood, metro station, bus stop, landmarks, etc., are all locational references but do not identify specific locations. Personal addresses are normally only shared by phone or private message. As Trably (2025, p. 347) writes, these local networks form a "local Internet where attachment to the area and mutual acquaintance frame interactions". For her, this "local Internet" is less an expression of a strictly geographical link than of "attachment to the territory through its associative and leisure practices and the mutual acquaintance that characterizes it (...)" (Trably 2024, p. 388). Participants must navigate between two types of hybridized mutual acquaintance: the concrete acquaintance of people who meet in the same places, and the digital acquaintance of local information exchanges. But here again, we find it difficult to separate attachment to a territory from the material and concrete topographical space formed by streets, squares, buildings, fields and forests, which concretely support local activities.

10 Since 2018 in the United States, the hyperlocal neighbors network has been connecting all owners of connected video doorbells from Amazon subsidiary Ring, enabling them to remotely monitor their doorsteps, send alerts and share videos of suspicious activity with neighbors and police. Visual surveillance is becoming natively geospatial.

While precise geolocation is not necessary on Facebook and Nextdoor, it should be remembered that both companies' business models are based on selling user data and targeted advertising. They have users' IP addresses, email addresses and phone numbers, as well as their geographical location at the neighborhood level and, in the case of Nextdoor, even their postal address. User profiles are all the more valuable to advertisers because they include details of daily activities in the living space. There is a clear contrast between the global scale of the companies that manage these networks and the "localized self" (Trably 2024, p. 375) that they help to generate among their users. These users devote time and energy to building collective local resources that are also useful to isolated individuals or those who have recently arrived in the area. As Trably (2025) notes, the movement calling for people to leave large local commercial networks in favor of tools that are less predatory in terms of data does not currently offer an alternative solution for people who, for a variety of reasons, depend on these large local social network operators.

1.5. Conclusion

In this chapter, we have described geodigitalization as a continuous process. However, certain stages of this process are marked by sociotechnical milestones linked to a series of innovations that are inseparable from companies marketing them, groups disseminating them and users finding uses for them – or not. The utopian project that inspires it is the creation of a computer-simulated replica of the material and mental world. Its purpose and consequences must be questioned. Nevertheless, geodigitalization also takes the form of multiple geospatial operations that change individuals' perception of reality, establish new regulation through the space of interactions and social behaviors and also result in concrete changes in the material environment. However, what we have described concerns only part of the information collected. Alongside computers and smartphones, a growing volume of information comes from what is known as the Internet of Things, that is, all those geolocated sensors that collect various data on their immediate environment: connected speakers, surveillance cameras, environmental sensors, access terminals to buildings and transportation, etc. In addition to all these static devices that directly inscribe computer code into space (Kitchin and Dodge 2011), there are also mobile sensors built into phones and connected vehicles. That said, the informational net that geodigitalization casts over reality is far from being seamless and homogeneous. Furthermore, its density varies greatly depending on the territory (Noucher 2023). It is more like a superimposition of irregular and incomplete grids, with different resolutions. However, mobilizing and combining the grids produced by geodigitalization requires specific engineering expertise, which is only available to large public bodies (and intelligence agencies), as well as large international digital information companies, which also have access to their gigantic proprietary databases. This raises significant concerns for an effective political and democratic regulation.

1.6. References

Arcaute, E., Barthelemy, M., Batty, M., Caldarelli, G., Gershenson, C., Helbing, D., Moreno, Y., Ramasco, J.J., Rozenblat, C., Sánchez, A. (2021). Future cities: Why digital twins need to take complexity science on board [Online]. Available at: https://www.researchgate.net/publication/354446988_Future_Cities_Why_Digital_Twins_Need_to_Take_Complexity_ Science_on_Board.

Auxier, B. and Anderson, M. (2021). Social media use in 2021. Pew Research Center [Online]. Available at: https://www.pewresearch.org/internet/2021/04/07/social-media-use-in-2021/.

Banque des territoires. (2021). *Miroir, miroir... : le jumeau numérique du territoire.* [Online]. Available at: https://www.banquedesterritoires.fr/miroir-miroir-le-jumeau-numerique-du-territoire-0.

Battiston, A., Napoli, L., Bajardi, P., Panisson, A., Perotti, A., Szell, M., Schifanella, R. (2023). Revealing the determinants of gender inequality in urban cycling with large-scale data. *EPJ Data Science*, 12(1), 9. doi: 10.1140/epjds/s13688-023-00385-7.

Beaude, B. (2012). *Internet, changer l'espace, changer la société.* FYP éditions, Limoges.

Besse, J.-M. and Tiberghien, G.A. (2017). *Opérations cartographiques.* Actes Sud/École nationale supérieure de paysage, Arles.

Boudreau, C. and Caron, D.J. (2022). Le déploiement de la surveillance épidémiologique automatisée du COVID-19 : une analyse comparée de la Corée du Sud et du Canada. *Canadian Public Administration*, 65(2), 261–277. doi: 10.1111/capa.12459.

Brachmann, E. and Prisacariu, V.A. (2024). Building a large geospatial model to achieve spatial intelligence [Online]. Available at: https://nianticlabs.com/news/largegeospatialmodel/.

Brown, M., Sanderson, Z., Graham, S., Kim, M., Tucker, J., Messing, S. (2024). Digital town square? Nextdoor's offline contexts and online discourse. *Journal of Quantitative Description: Digital Media*, 4. doi: 10.51685/jqd.2024.icwsm.2.

Chrisman, N.R. (2006). *Charting the Unknown: How Computer Mapping at Harvard Became GIS.* ESRI Press, Redlands, CA.

Colley, A., Thebault-Spieker, J., Lin, A.Y., Degraen, D., Fischman, B., Häkkilä, J., Kuehl, K., Nisi, V., Nunes, N.J., Wenig, N., et al. (2017). The geography of PokéMon GO: Beneficial and problematic effects on places and movement. In *CHI '17: Proceedings of the 2017 CHI Conference on Human Factors in Computing Systems.* doi: 10.1145/3025453.3025495.

Courmont, A. (2018). Plateforme, big data et recomposition du gouvernement urbain. *Revue française de sociologie*, 59(3), 423–449. doi: 10.3917/rfs.593.0423.

Crumplin, W.W. (2007). Geographic information systems as media and society: Does GIS wear a white or black Stetson? *Cartographica: The International Journal for Geographic Information and Geovisualization*, 42(1), 56–86. doi: 10.3138/carto-v42-1-065.

Cunty, C., Mathian, H., Cartomouv, G. (2017). Les pratiques de cartographie animée pour représenter le changement. *Mappemonde. Revue trimestrielle sur l'image géographique et les formes du territoire*, 120. doi: 10.4000/mappemonde.2602.

Foresman, T.W. (1998). *The History of Geographic Information Systems: Perspectives from the Pioneers*. Prentice Hall PTR, Upper Saddle River, NJ.

Furno, N. (2016). Waze perturbe les quartiers résidentiels. *iGeneration* [Online]. Available at: https://www.igen.fr/app-store/2016/06/waze-perturbe-les-quartiers-residentiels-96058.

George, P. (1978). *Les méthodes de la géographie*. Presses Universitaires de France, Paris.

Goodchild, M.F. (2007). Citizens as sensors: The world of volunteered geography. *GeoJournal*, 69(4), 211–221.

Groupe Dupont (1994). *GÉOPOINT 94. SIG, analyse spatiale et aménagement*. Université d'Avignon, Avignon.

Guérois, M. and Madelin, M. (2024). Heterogeneous data integration and Geoweb cartographic representations. In *Handling and Mapping Geographic Information*, Cunty, C. and Mathian, H. (eds). ISTE Ltd, London and Wiley, Hoboken, NJ. doi: 10.51926/ISTE.9161.ch2.

Guillaud, H. (2021). Application de suivi de contact : quel bilan ? InternetActu.net [Online]. Available at: https://www.internetactu.net/2021/09/20/application-de-suivi-de-contact-quel-bilan/.

Hansen, L., Arnesen, P., Graupner, S.-T., Lindkvist, H., Leonardi, J., Al Fahel, R., Andersson, K. (2021). Current state of the art and use case description on geofencing for traffic management. *GeoSence* [Online]. Available at: https://sintef.brage.unit.no/sintef-xmlui/handle/11250/2826636.

Herdağdelen, A. and Adamic, L. (2023). The geography of Facebook groups in the United States. *Proceedings of the International AAAI Conference on Web and Social Media*. 17, 351–362 [Online]. Available at: https://ojs.aaai.org/index.php/ICWSM/article/view/22151.

Hill, L. (2006). *Georeferencing. The Geographic Associations of Information*. MIT Press, Cambridge, MA.

Iqbal, W., Ghafouri, V., Tyson, G., Suarez-Tangil, G., Castro, I. (2023). Lady and the tramp nextdoor: Online manifestations of real-world inequalities in the nextdoor social network. *Proceedings of the Seventeenth International AAAI Conference on Web and Social Media*, 17, 399–410 [Online]. Available at: https://ojs.aaai.org/index.php/ICWSM/article/view/22155.

Joliveau, T. (1996). Gérer l'environnement avec des S.I.G. Mais qu'est-ce qu'un S.I.G. ? *Revue de Géographie de Lyon*, 71(2/96), 101–110.

Joliveau, T. (2004). Géomatique et gestion environnementale du territoire. Recherches sur un usage géographique des SIG. Thesis, Université de Rouen [Online]. Available at: https://hal.science/tel-04559188v1.

Joliveau, T. (2007). Géomatique et géonumérisation. Monde géonumérique [Online]. Available at: https://mondegeonumerique.wordpress.com/geomatique-et-cie/geomatique-et-geonumerisation/.

Joliveau, T. (2010). La géographie et la géomatique au crible de la néogéographie. *Tracés*, 3(HS-10), 227–239.

Joliveau, T. (2011). Le géoweb, un nouveau défi pour les bases de données géographiques. *L'Espace Géographique*, 40(2), 151–163.

Joliveau, T. (2012). Les espaces réels dans le jeu vidéo. Contribution à une approche géographique des espaces vidéoludiques. In *Espace et temps des jeux vidéo*, Rufat, S., Minassian, H.T., Coavoux, S. (eds). Questions Théoriques, Paris.

Kebs, B. (2024). The global surveillance free-for-all in mobile ad data – Krebs on security [Online]. Available at: https://krebsonsecurity.com/2024/10/the-global-surveillance-free-for-all-in-mobile-ad-data/.

Kennedy, L. and Coelho, M. (2022). Security, suspicion, and surveillance? There's an app for that. *Surveillance & Society*, 20(2), 127–141. doi: 10.24908/ss.v20i2.14536.

Kitchin, R. and Dodge, M. (2011). *Code/Space: Software and Everyday Life*. MIT Press, Cambridge, MA.

Kurwa, R. (2019). Building the digitally gated community: The case of nextdoor. *Surveillance & Society*, 17(1/2), 111–117. doi: 10.24908/ss.v17i1/2.12927.

Lucas, J.-F. (2022). Imaginaires et traditions des jumeaux numériques urbains. In *La renaissance industrielle*, Musso, P. (ed.). Éditions Manucius, Paris [Online]. Available at: https://hal.science/hal-03816032.

Mangeart, T. (2023). Plateformisation de la mobilité et territorialisation des plateformes : le cas du free-floating en Île-de-France. *Netcom. Réseaux, communication et territoires*, 37(3/4). doi: 10.4000/vyjj.

Mericskay, B. (2016). La cartographie à l'heure du Géoweb : retour sur les nouveaux modes de représentation spatiale des données numériques. *Cartes & géomatique*, 37, 229–230.

Nextdoor (2024). Nextdoor year in review: What Canadian neighbours were talking about in 2024. *About Nextdoor* [Online]. Available at: https://about.nextdoor.com/press-releases/what-canadian-neighbours-talked-about-in-2024/.

Noucher, M. (2017). *Les petites cartes du web. Analyse critique des nouvelles fabriques cartographiques*. Éditions Rue d'Ulm, Paris [Online]. Available at: http://www.presses.ens.fr/481-actes-de-la-recherche-a-l-ens-petites-cartes-du-web-les.html.

Noucher, M. (2023). *Blancs des cartes et boîtes noires algorithmiques*. CNRS Éditions, Paris.

Nova, N. (2009). *Les médias géolocalisés*. FYP Éditions, Limoges.

Odobašić, D., Medak, D., Hardegen, M. (2013). Gamification of geographic data collection. In *GI_Forum 2013, Creating the GISociety – Conference Proceedings*. Verlag der Österreichischen Akademie der Wissenschaften. doi: 10.1553/giscience2013s328.

Payne, W. (2017). Welcome to the polygon: Contested digital neighborhoods and spatialized segregation on nextdoor. *Computational Culture*, 6 [Online]. Available at: http://computationalculture.net/welcome-to-the-polygon-contested-digital-neighborhoods-and-spatialized-segregation-on-nextdoor/.

Philipponneau, M. (1960). *Géographie et action : introduction à la géographie appliquée*. Librairie Armand Colin, Paris.

Rimbert, S. (1989). G.I.S. ou pas ? *Mappemonde*, 1, 1–3.

Scharl, A. and Tochtermann, K. (2007). *The Geospatial Web: How Geobrowsers, Social Software and the Web 2.0 are Shaping the Network Society*. Springer, London.

Schuurman, N. (2000). Trouble in the heartland: GIS and its critics in the 1990s. *Progress in Human Geography*, 24(4), 569–590. doi: 10.1191/030913200100189111.

Sénécat, A., Untersinger, M., Imbach, R. (2025). Comment des applications grand public facilitent le pistage des utilisateurs à leur insu. *Le Monde* [Online]. Available at: https://www.lemonde.fr/pixels/article/2025/01/15/comment-des-applications-grand-public-facilitent-le-pistage-des-utilisateurs-a-leur-insu_6498708_4408996.html.

Sui, D.Z. and Goodchild, M.F. (2011). The convergence of GIS and social media: Challenges for GIScience. *International Journal of Geographical Information Science*, 25(11), 1737–1748. doi: 10.1080/13658816.2011.604636.

Sui, D.Z. and Goodchild, M.F. (2003). A tetradic analysis of GIS and society using McLuhan's law of the media. *The Canadian Geographer/Le Géographe canadien*, 47(1), 5–17. doi: 10.1111/1541-0064.02e08.

Trably, L. (2024). Se dévoiler sur les réseaux. PhD Thesis, Université Paris Cité, Paris [Online]. Available at: https://hal.science/tel-04880315.

Trably, L. (2025). Mais que font encore les internautes sur Facebook ? AOC media – Analyse Opinion Critique.

Turkle, S. (1995). *Life on the Screen: Identity in the Age of the Internet*. Simon & Schuster, New York [Online]. Available at: http://archive.org/details/lifeonscreeniden0000turk.

Turner, A.J. (2006). *Introduction to Neogeography*. O'Reilly Media, Inc., Farnham.

Untersinger, M. (2024, October 27, 2024). Strava, une histoire émaillée de failles de sécurité. *Le Monde*.

Vial, S. (2014). Contre le virtuel : une déconstruction. *MEI – Médiation et information*, 37, 177–188.

Zahnow, R., Verrier, J., Hames, S., Corcoran, J. (2024). Mapping and measuring neighbourhood social media groups. The case of Facebook. *Applied Geography*, 172, 103415. doi: 10.1016/j.apgeog.2024.103415.

Zook, M.A. and Graham, M. (2007). Mapping DigiPlace: Geocoded Internet data and the representation of place. *Environment and Planning B: Planning and Design*, 34(3), 466–482. doi: 10.1068/b3311.

2

Algorithmics at the Heart of Web-based Geographic Data Production

Teriitutea QUESNOT
UMR ESPACE, Université Côte d'Azur Nice,
Institut Universitaire de France Paris, France

2.1. The genesis of geographic "data"

"Big data", "small data", "open data", "data science", "data papers", "dataviz", "datafication", and "data days" – the list of expressions that misuse the term *data* keeps growing. Its usage has intensified over the past decade (2015–2025), fueled by the hyper-mediatization of massive datasets produced and owned by Big Tech companies. In the so-called "data era", geography and its associated disciplines – cartography, geomatics and related fields – are often portrayed as occupying a pivotal position, echoing the well-known, though controversial, claim that "80% of all information contains some geospatial reference". This oft-repeated assertion originates from a paper Williams presented at the 1987 URISA conference, where the then-director of the Alachua County Regional Information Center (USA) estimated that 80% of the information needed to manage his county had a spatial dimension – that is, location-based information (Williams 1987, cited in Garson and Biggs 1992).

The few studies that have examined this estimate suggest that the actual figure is closer to 60% (Hahmann and Burghardt 2013). Beyond this numerical discrepancy,

Geospatial Technologies and Society,
coordinated by Thierry JOLIVEAU and Matthieu NOUCHER.

what deserves attention is the semantic shift that has occurred over the past two decades. Today, the conversation is no longer about 80% of geographic *information*, but rather 80% of geographic *data*[1]. This distinction is not trivial. Beyond highlighting the current data hype, it underscores a significant point: contrary to its etymological implication, data are not raw materials waiting to be collected and analyzed to yield information (and, by extension, knowledge). Data are never raw – they are constructions, fabricated, (re)arranged and continuously (re)combined (Kitchin 2014). Confusing data with information is, in a sense, acknowledging the absence of objectivity in what we conveniently call "data".

On the web, all digital data – without exception – are shaped directly or indirectly by algorithms, themselves designed by engineers (and increasingly by AI systems). But what about geographic data[2] produced online? More fundamentally, what do we actually mean by "geographic data"? What does the qualifier *geographic* – often conflated with *spatial* or *geospatial* – actually refer to? Does its geographic nature automatically imply a particular form of objectivation, grounded in protocols of conceptualization and collection that aim to verify and document potential biases relative to specific objectives – impartiality, factual accuracy and measurement quality (Daston and Galison 2012)? These questions are legitimate and deserve further clarification.

In its broadest sense, a dataset is considered geographic (or spatial) when it refers to one or more objects, individuals or phenomena located on Earth (Aschan-Leygonie et al. 2019).

Outside of the web sphere, geographic data are obtained through a variety of channels: physical measurement instruments (for instance, a sensor designed to record temperature), satellite imagery (for example, to produce a land-cover map), field surveys (such as population censuses) or Internet of Things (IoT) sensors (for example, connected devices measuring urban pollution). In all of these cases, objectivity is never absolute: physical instruments must be calibrated in specific ways, satellite images are produced under particular conditions (date of acquisition, spatial resolution, etc.), field surveys are governed by prior methodological choices (selection criteria, sampling strategy and so forth), and IoT sensors are constrained by technical and location factors that affect their readings (for instance, pollution sensors may be influenced by their proximity to emission sources or by local microclimatic variations). Data – whether geographic or not – are always collected for specific

1 See the following discussion on the StackExchange Q&A network: https://gis.stackexchange.com/questions/18838/80-of-data-has-a-spatial-component-says-who.

2 Since this chapter focuses on data generated on the web, the word "data" will henceforth implicitly refer to digital data.

purposes and according to predefined protocols. Their validity and relevance therefore depend on the rigor and transparency of those very protocols, which should be validated, described and communicated to allow for critical scrutiny. Yet such conditions are not always met in practice.

Furthermore, geographic data differ from other categories of data in that they contain location information. This spatial component can take various forms – from simple association with named geographic entities (toponyms) to more or less precise coordinates (x/y or latitude/longitude). In practice, whenever a location must be assigned, this process raises significant considerations, as it is not neutral. It entails a series of choices, most notably the selection of a spatial reference system, which requires defining several parameters such as the ellipsoid, geodetic datum and prime meridian. In the case of projected coordinate systems, this is further complicated by the choice of a cartographic projection, which inevitably introduces distortions. The selection of all these parameters – guided by criteria such as the scale of the study area or the nature of the features being located – ultimately depends on the decision of the geographic data producer(s). These choices are not devoid of subjectivity and thus have significant implications for how geographic space is represented and interpreted.

Therefore, geographic data, while aiming to describe the position of objects and phenomena on Earth, necessarily do so through a series of abstractions. These abstractions situate entities within a spatial reference system that, although mathematical in nature, remains a human construction. Far from being any less real, such abstractions are essential tools for apprehending and managing the complexity of the physical environment. Over the decades, they have evolved alongside technological progress – particularly with the emergence of tools designed to handle and analyze geographic data, namely, Geographic Information Systems (GIS). These technological advances have facilitated the processing of geographic information but have simultaneously deepened and complicated the layers of abstraction inherent in the production and manipulation of geographic data. Undeniably, these developments have not only transformed our ability to represent and analyze space, but have also profoundly reshaped our very conceptualization of geographic information.

The evolution of GIS perfectly illustrates this trajectory (see Chrisman 2004) – from the first attempts to digitize geographic information to today's data-sharing platforms, which store ever-increasing volumes of geographic data. These datasets (and their metadata) have become increasingly open:

1) catalogued on Spatial Data Infrastructures (SDI);

2) directly downloadable;

3) generally accessible through web services such as Web Map Services (WMS) and Web Feature Services (WFS).

In many respects, this ecosystem and its web-oriented vocabulary reinforce the perception of geographic data as data in the strict sense – stored in repositories, waiting to be (re)collected for analytical and/or communicational purposes. Should we then abandon the term *data* when referring to these digital renderings of the physical environment? Despite their evident lack of objectivity, it remains appropriate to retain the expression *geographic data*. Beyond convenience of use, objectivity here may be understood as a claim to reliability (Longino 1990). Much like the content of Ptolemy's seminal treatise *Cosmographia* (150 CE), geographic data disseminated through SDIs are intended to serve as reliable representations of the environment and are suitable for reuse.

All of these technological advances are part of a broader quest to represent the world ever more accurately. With this in mind, the concept of "digital twin" has recently emerged as a new step in this progression. Extending the ambition of traditional GIS, the digital twin seeks to create a more comprehensive and dynamic virtual replica of the physical environment. However, despite its apparent sophistication, it remains subject to the same constraints of abstraction and simplification as GIS. While it integrates real-time data and enables more complex simulations, it cannot escape the limitations inherent in any modeling. However advanced it may be, a digital twin remains a partial and selective representation of reality – one influenced by configuration parameters (e.g. geometric detail, data-update frequency and so forth), the technologies employed and the goals that guided its creation. The pursuit of a perfectly mimetic representation of the world thus remains an ideal that continues to drive the development of geographic data but can never be fully attained.

The constructed nature of geographic data raises important ethical and societal issues, echoing the pioneering work of Harley (1989) on critical cartography (and critical GIS) (see also Crampton 2001). The choices made during the collection, processing and representation of geographic data can shape public policies, spatial planning and even how individuals perceive their surroundings. Moreover, unequal access to these data, and to the tools required to manipulate them, can exacerbate existing disparities among social groups and/or regions. Finally, geographic data represent only one component of the broader landscape of geospatial information on the web. They coexist and interact with other forms of geographic data – notably *capta* (geolocated content from social media) and *quaesita* (real-time information designed to influence spatial behaviors) – whose definitions and implications will be explored in the following sections of this chapter.

2.2. Social web platforms and the proliferation of "capta"

The arrival of Web 2.0 was marked by the development of participatory platforms such as Wikipedia. In the field of cartography, and more generally geographic information, OpenStreetMap (OSM) remains by far the most emblematic community mapping project on the web (see Chapter 3). The digitization of the physical environment, previously reserved for public authorities, was opened up to citizens. Three years after OSM's launch, the number of volunteer mappers exploded, growing from a few hundred to nearly 45,000 (Haklay and Weber 2008). Today, the platform has over ten million registered users[3]. OSM geographic data available are used by numerous organizations, from Big Tech companies (like Meta) to NGOs (such as CartONG) and government institutions, including Statistics Canada. As a matter of fact, OSM embodies three major trends:

1) Open data, with all OSM data distributed under the ODbL license[4].

2) Geocommons, which refer to geographic databases governed according to the commons principle (Ostrom 1990).

3) Volunteered Geographic Information (VGI[5]), as theorized by Goodchild (2007).

Data dedicated to digitizing the physical environment are not the only forms of geographic data produced on the web. Web 2.0 (O'Reilly 2005) notably enabled the emergence of digital social networks. Meta, Twitter (X), Foursquare, Instagram and TikTok are now essential parts of the social web. They generate a considerable volume of "data" with geographic characteristics that have nothing to do with OSM data. They are closed (proprietary), do not follow the commons principle and do not

3 OSM's official statistics are regularly updated on the following web page: https://wiki.openstreetmap.org/wiki/Stats.

4 For more information on the Open Data Commons Open Database License, see https://opendatacommons.org/licenses/odbl/.

5 The concept of VGI refers to the creation, collection, and dissemination of geographic data by volunteer citizens rather than professional organizations. This term, introduced by Goodchild (2007), describes a phenomenon where users, often non-experts, become true sensors producing geographic information.

fall within the sphere of VGI[6]. The following paragraphs will demonstrate that these "geosocial data" cannot properly be called "data", as they reflect no geographic reality whatsoever – neither physical nor social, and certainly not "geosocial".

Historically, the first location-based social networks (LBSNs) – Brightkite and Gowalla – appeared in 2007, following in the footsteps of Facebook (now Meta), which had been launched three years earlier. In concrete terms, LBSNs revolve around the practice of "checking in", which consists of sharing our location to all members of our network. Brightkite and Gowalla were acquired by Limbo in 2009 and Facebook in 2011, effectively clearing the way for Foursquare, the only real competitor at the time, to establish itself as the reference in social geolocation. Unlike its predecessors, Foursquare innovated by "gamifying" the act of publishing a check-in through the "Mayors" system. To encourage users to publish check-ins and register new locations, the platform distributed badges and titles each week to users who accumulated the most check-ins. Depending on which venue they were mayor of, users could receive various perks (discounts, free admission, etc.).

And the formula worked – at least for a few years.

In fact, obtaining these rewards ranked among the primary motivations for using Foursquare in the survey conducted by Lindqvist et al. (2011). Check-ins quickly became a means of communication centered on places and their notoriety (Quesnot and Roche 2015a, 2015b), with users having an array of indicators such as "likes", ratings and comments to express their opinions and feelings. Traditional digital social networks (Facebook, X, etc.) subsequently integrated the basic features of LBSNs, particularly check-ins. This widespread adoption considerably expanded the reach and impact of geolocated data on social web platforms. This avalanche of content from other digital social networks unsurprisingly attracted the attention of the scientific community. In the 2010s, numerous researchers seized upon it to extract the sense of place and understand the complexity of human spatiality (Lussault 2007) (see the Livehoods project by Cranshaw et al. (2012) as well as the Social Media Ingestor by Stefanidis et al. (2013)).

In a position study, we already explained that "geosocial data" held little interest for geographers for three reasons (Quesnot 2016a). First, these data are not exhaustive, contrary to what some think or claim. The data retrieved are partial and biased: users of these networks are not representative of the general population, and the data returned by APIs are systematically filtered by algorithms about which we

6 Fischer (2012) introduced the term "iVGI" (Involuntary Geographic Information) to distinguish geographic data involuntarily generated by users of social web platforms from those produced within the framework of traditional VGI.

know nothing (or almost nothing). The same query sent to an API may return different results, and their terms of use can change overnight, which clearly contradicts the principle of reproducibility in conventional scientific methodology. Second, these "data" are completely acontextual. They reveal nothing about the context in which they were produced. We know, for example, that a difficult-to-estimate number of tweets are generated automatically by bots, and that a check-in does not necessarily equate to visiting a place[7] (Rost et al. 2013). Third, "geosocial data", like the majority of geotagged digital content, are manufactured by digital giants to generate profits. They are not a direct window into our daily spatialities. Their value, artificially created to make them marketable, is intrinsically linked to their volume. Unlike natural resources, it is not the scarcity of the "data" that creates value here, but its abundance. It is therefore essential for the Big Tech companies to saturate the web with these "geosocial data" and other geotagged content (check-ins, location tags automatically attached to photos and videos, etc.).

Therefore, it appears incorrect to consider emanations from social web platforms as geographic *data* per se, as their primary purpose is communication. They are not produced by Internet users with the intent of being processed geographically[8] – their geographic nature is most often limited to a geolocation tag (x/y or latitude/longitude) attached to digital content that is not necessarily spatial. While geographic data aim to be a faithful digitization of the environment, these emanations are by no means reliable indicators of the ideational and emotional dimension of human spatiality (Quesnot 2016a). Under these conditions, and in agreement with Rzeszewski (2016), the term "geographic *capta*" would be more appropriate for referring to this family of "data".

For Drucker (2011), what we commonly call "data" does not align with the original meaning of the word *data*, which refers to something assumed to be given (in the sense of "served up"). According to Drucker, all "data" are in fact "capta" – that is, things obtained with the aim of producing knowledge. Fundamentally, Drucker's position is legitimate: in science, as elsewhere, there are no things (quantities or otherwise) that are truly served up ex nihilo. All of the "data" we manipulate are constructions and are therefore necessarily partial. However, systematically replacing the term "data" with "capta" is debatable, particularly in the

7 The publication of a check-in is free and is not subject to "geographic verification" (i.e. verification that the location declared matches the user's position at the time of publication).

8 Obviously, certain forms of capta can be analyzed retrospectively to extract geographic information. For example, visible elements in a photograph (signs, road signs, monuments, etc.) can be studied to infer the location of the photographed scene.

realm of geographic data from the web. As Lavin (2021) points out, the etymology and historical usage of the term "data" are more complex than a simple opposition between "given" and "taken". The history of the word "data" is closely linked to the publication and translations of Euclid's work, where *data* originally referred to the quantities used to solve a mathematical problem, as opposed to *quaesita*, the quantities being sought. In fact, two distinct conceptions of *data* coexist: a geometric reasoning approach on the one hand (Euclid), and an empirical approach to transcribing and representing physical objects on the other hand. Lavin thus advocates for a nuanced view of the term "data" in which data result from a complex process involving collection, assembly and recording by individuals or their instruments. Their structure stems both from the methodology of observation and from the interpretive choices made during their genesis. This conception of "data as situated knowledge" highlights their dual nature: they are constructed from structured observations of phenomena while incorporating the interpretations anticipated during their collection. This perspective ultimately recognizes the inherently empirical and interpretive dimension of *data*.

Although Lavin critiques Drucker's approach and advocates for more thoughtful usage of the term "data", the distinction between *data* and *capta* remains relevant in certain contexts, notably that of geographic data from the social web. Indeed, without necessarily endorsing Drucker's proposal to replace "data" with "capta", we can consider these terms as part of a continuum reflecting different degrees of intervention and appropriation in the process of data collection and use.

It is therefore essential to distinguish between physical environment digitization data and "geographic data" from the social web, without strictly opposing them. The former, though socially constructed, aim to translate as faithfully as possible the material dimension of the environment. The latter are products of social web platforms whose business model relies on capturing and monetizing users' data. The use of the term *capta* to qualify the latter then takes on its full meaning, not only in the sense of "taking", but also in that of "capturing". Indeed, social web platforms no longer simply generate data as by-products of their services. They have, to some extent, reversed their operational logic, organizing these services with the explicit goal of encouraging their users to produce ever more data.

This approach consists of nothing less than capturing (almost by force) citizen-generated digital content, transforming it, then isolating it to artificially valorize it for profit. Thus, the difference between geographic *data* and *capta* lies less in their intrinsic nature than in their mode of collection and distribution – social web data being collected often without users' awareness, and their distribution remaining controlled by the financial interests of Big Tech platforms.

2.3. From geospatial footprints to geographic "quaesita"

There is a specific category of "geographic data" that has not yet been discussed and which is generally grouped under the term "traces". In the context of digital mapping, Noucher (2015) considers traces as "a set of information, spatialized or not, left voluntarily or involuntarily by an individual on the Internet and which can be subject to cartographic processing" (p. 215). He identifies four types of traces based on two axes (intentional/unintentional traces and individual-centered/individual data aggregation mapping):

1) those revealed by citizen-sensors (e.g. routes uploaded by users of the application RunKeeper);

2) those produced from GPS sensors (typically, naturalist observations within citizen science);

3) those that come from cyber-traces, such as Google histories that allow for the creation of personalized maps;

4) those extrapolated from digital footprints, such as geolocation data from our smartphones.

Mericksay et al. (2018) propose a similar, albeit slightly more detailed, typology:

1) flow traces (movement) and mark traces (crystallization of movement);

2) voluntary traces (VGI) and involuntary traces (e.g. when a user unconsciously uploads geolocation data from their smartphone);

3) implicitly geolocated traces, such as geolocation tags inferred from various processes like geocoding or geoparsing, and explicitly geolocated traces (assignment of a geolocation tag during the creation of digital content);

4) individual traces and aggregated traces.

An alternative typology can be formalized by drawing on Galinon-Mélénec's (2013a, 2013b) *Homme-trace* (Trace-Man) paradigm. The author posits that our interactions with the environment are inscribed in our body in extenso – the brain itself being affected – in the form of sign-traces, and that through feedback, we inscribe other sign-traces in the environment. In this theoretical framework, the reality corresponds to an entanglement of sign-traces far too complex to be fully apprehended through the five senses – which is why humans would seek to simplify it by introducing discontinuities encapsulated in the form of words, concepts, notions and images.

As a starting point, let us revisit the definition of "trace". For Galinon-Mélénec (2015), to speak of a trace is to assume that something deposited in the past has been preserved in the present. A trace therefore refers to a deposit left – voluntarily or not – in an environment during a process that, unlike the trace, is inherently dynamic. This deposit is a part of a whole: a past situation that we seek to reconstruct here and now (otherwise, the notion of trace would lose all meaning). Therefore, a trace does not exist by itself. It is something necessarily signified by the observer who attempts to reconstruct a past situation according to their own inferential schema. Galinon-Mélénec insists on the term "deposit" because, for her, a trace does not necessarily imply an inscription in matter, unlike an imprint. In other words, the imprint is nothing more than a subset of the trace that distinguishes itself by being inscribed more or less durably in "matter". This difference is crucial because it implies difficulty, or in some cases, impossibility of erasure. Based on these theoretical considerations, it is logical to assert that all digital content constitutes an imprint, since it is stored (recorded) in binary form (0-1) in a computer system.

Consequently, all "data" on the web – whether geographic or not – represent full-fledged digital footprints, whether they are "cyber-traces" or information uploaded by citizen-sensors and GPS chips. The same applies to physical environment digitization data and geographic *capta*. To this super-category, it is essential to add an additional distinction based on a fundamental yet (too) often overlooked implication: that of the body. Indeed, geographic *data* and *capta* are inherently disembodied, their production never requiring active engagement of the body[9]. In absolute terms, check-ins and geotagged digital content do not necessarily require movement through the physical environment. Often, users use these posts to distinguish themselves from the collective through the places they claim[10] to visit. In contrast, all geolocation information uploaded in real time to the web by the GPS chips of our mobile devices ipso facto reflect our movement through/in/with space. These "geospatial footprints" thus differ significantly from the digital footprints mentioned above in their experiential dimension[11]. They represent, so to speak, the digital inscription of movement on the web.

Galinon-Mélénec (2015) does not stop at the imprint and continues her reasoning by explaining the concept of "clue". According to her, the search for clues is

9 It is important to note that the production of *data* or *capta* may well involve bodily engagement, typically movement (e.g. a "reel" posted on Instagram). However, their creation does not depend on movement through the physical environment.

10 We use the verb "claim" because the user can easily claim that they are in the location without actually being physically present there.

11 At least, for geographic data that is produced off-site (e.g. from satellite images).

inseparable from the process of dividing up the continuum of reality. More specifically, an imprint can potentially become a clue when there is an observer seeking meaning. Like a trace, a clue does not exist on its own. It depends on the observer who, after incorporating a trace[12], must make a judgment to decide that it is a clue. This judgment is based on a pre-established schema that will be confirmed or refuted a posteriori, and it is precisely here that the concepts of signal-sign and trace-sign come into play. Reality, in its complexity, can only be partially apprehended through signs. Galinon-Mélénec (2013b) defines the signal-sign as a trace that captures an individual's attention, without this capture necessarily being conscious. This trace is then qualified as a trace-sign because it constitutes a deposit potentially accessible and recognizable in the present. Interactions between the individual and their environment continuously produce new trace-signs, which increases the complexity of their interpretation. Faced with this profusion of trace-signs, humans necessarily perform selection and interpretation. As Galinon-Mélénec points out, this interpretive process is strongly influenced by the individual's personal history and the sociocultural milieu from which they come. Each person develops their own way of apprehending and giving meaning to the trace-signs that surround them. Constantly searching for clues, humans are thus guided by signal-signs and produce in return a set of traces that in turn will play for themselves and/or for others the role of imprint, clue or signal-sign/trace-sign.

The semantics and logic of the *Homme-trace* paradigm prove useful for deciphering the nature of geospatial footprints and their corollary, grouped here under the term "geographic *quaesita*". The choice of the term *quaesita* to qualify these geographic data of a new kind is significant. As a reminder, *quaesita* designated in Euclid the quantities sought in a geometric problem, as opposed to *data*, which represented given quantities. By transposing this concept to modern geospatial technologies, this terminology underscores a fundamental characteristic: this new kind of data is not simply given (*data*) or captured (*capta*), but actively constructed and deployed to influence our spatial behaviors. Just as Euclidean *quaesita* represented the unknowns to be determined, geographic *quaesita* constitute the parameters that algorithms seek to optimize to shape our relationship to geographic space.

This novel conceptualization becomes particularly relevant when we examine how digital giants exploit our travels. Indeed, our movements through space are notably recorded with the aim of determining marketable profiles. In this spirit, the aggregation of our geospatial footprints is inevitable. Since movement can only be

12 Reality being a continuum, the trace in question is de facto assimilated by an individual, here in the cognitive sense of the term.

experienced – it is a matter of lived experience – it is simply impossible for us to encapsulate it in a computer system. The movements we see unfolding in "real time" on our navigation applications (Google Maps, Apple Maps, etc.) are in fact only a posteriori reconstructions of geospatial footprints captured at very short time intervals or extrapolated[13]. With our routes reconstituted and recorded over time, the algorithms of these applications are able of drawing a "map" of our spatial behaviors, both individual and collective. These representations are similar to inferential patterns used to anticipate our movements. When we navigate through the physical environment, these same "learning algorithms" (machine/deep learning) are constantly searching for clues that will confirm or, if necessary, refute these inferential schemas. On the lookout for the slightest signal, they adjust their "behavior" (almost) in real time. They are programmed to do so. Unlike analytical tools that rely solely on traditional statistical methods, these Big Data bots are also keen on outliers. This ability to rapidly integrate unexpected data and modify movement suggestions accordingly is a key characteristic of geographic *quaesita*. A sudden change of direction, and the system immediately suggests an alternative route. An unexpected stop during your walk, and the navigation application deduces that you probably visited the place located near that stop. A disguised confirmation request like "What did you think of this place?" is then sent to you. You are then "free" to leave yet another digital footprint (rating, comment, etc.) on the web. This continuous cycle of collection, interpretation and influence on spatial behaviors clearly illustrates the distinctive nature of geographic *quaesita* compared to simple traces or *capta*.

Moreover, the *Homme-trace* paradigm posits that we are not only producers but also assimilators of trace-signs. While attention is often focused on the (geo)digital footprints we inscribe here and there on the web, it appears equally important to carefully examine how the web imprints itself on us. Because to some extent, our brain, and by extension our memory, bears "cognitive traces" from our interactions with web platforms, and more generally with the digital universe. In fact, few Google users have never been influenced by the ratings given to commercial establishments or by the traffic predictions anticipated by Google Maps. However, this influence is no longer limited to these "simple" interactions. Thanks to behavioral maps constructed from our digital footprints (*capta*) and geospatial footprints (traces of our daily movements), web giants are now capable of inducing – very often subtly – a change in our daily spatialities by using geographic *quaesita*. Inducing behavior in a user (in other words, solving the mathematical problem, here applied to our movements and interactions with the environment) by deliberately introducing a parameter (i.e. a *quaesita*) based on existing parameters (i.e. *data*) is

13 Extrapolation is particularly useful for compensating for weak GPS signals.

not new, and even less so in the web sphere, since this is the modus operandi of algorithmic governmentality (Rouvroy and Berns 2013). Amazon excels at this game through its personalization algorithms which, using hyper-targeted notifications, encourage its customers to acquire (ever more) products adapted to their consumption habits.

Geographic *quaesita* differ from "ordinary" *quaesita* because they do not merely influence our consumption choices: they act concretely on how we physically interact through/in/with the immediate environment. The nudges[14] deployed by Google Maps thus belong to the category of geographic *quaesita*. For example, the platform now encourages drivers to adopt "eco-friendly" behavior by proposing a low CO_2 emission route. Like CityMapper, the application also ensures that the user gets off at the tram stop shown on their route by emitting a vibration. Finally, showing how busy places are in real time, Google regulates travel by encouraging (or discouraging) users from going into certain stores or subway cars when they check the mapping platform. These examples illustrate how geographic *quaesita* operate at the interface between the digital and the physical, and directly influence our interactions with our immediate environment.

Before entering into the deconstruction of (geo)digital footprints, several important points should be emphasized. First, we must be aware of the overquantification/disqualification operated by web giants with regard to our daily spatialities. Our relationships to places and the meaning(s) we give them are effectively reduced to attendance indexes and ratings (Quesnot 2016a, 2016b). Our way of being in the world, of "doing with" space (Lussault 2007), is here interpreted only in terms of flows (Quesnot and Roche 2020), with algorithms focusing solely on our way of doing on space. They do not take into consideration the sensory dimension of our experience of places because they are simply incapable of it. The space they shape, refine and continuously update using our (geo)digital footprints inscribed here and there is Cartesian per se. It is a cold space with stable Euclidean metrics, devoid of all sensibility. Also, it is precisely in this context that geographic *quaesita* act as intermediaries between this quantified (non-experiential) space and our own (experiential) spatialities. This same space is, as Rouvroy (2017a) rightly asserts, a space of speculation. This is not an attempt to construct a digital twin of the physical environment – geographic *data* creators are charged with modeling that. No. The speculative space referred to here is sketched from our spatial behaviors that GAFAM seek at all costs to anticipate but also, and above all, to regulate. It is a

14 Nudges are "soft" influence techniques aimed at modifying individuals' behavior in predictable ways, without resorting to prohibition or constraint.

space of rationalization that aims to neutralize as much as possible any deviation from pre-established inferential schemas.

Geographic *quaesita* therefore play a crucial role here, serving both as prediction tools and as means of influencing our spatial behaviors. The modification that Google made to its traffic prediction algorithms to deal with the extraordinary drop in traffic that occurred during the Covid-19 pandemic is a striking example of this. This unforeseen drop completely destabilized Google Maps' learning algorithms to the point that Google chose to ignore the vast majority of geospatial footprints inscribed by its users between 2019 and 2020[15]. Finally, the extreme porosity that persists between this digital space of speculation and our physical environment should be noted. The last example clearly supports this point. Because they control this speculative space, digital giants also have a stranglehold on our concrete and ideational relationships to the physical environment. This is therefore not only a space of neutralization. It also constitutes a space of pure potentialities (Rouvroy 2017b).

2.4. Deconstructing our (geo)digital footprints: pipe dreams and realities

Harley's (1989) deconstructionist practice aims to break with the positivist vision of "scientific" cartography that has been largely dominant since Ptolemy's seminal works (2nd-century CE). For the geographer and father of critical cartography, maps are presented as neutral and objective representations of reality – more precisely, of our physical environment – when in fact they are not. Drawing largely on Foucauldian and Derridean thought, Harley textualizes the map based on the principle that the "knowledge" it encodes necessarily contains an implicit power masked by its apparent objectivity. According to him, this power – which must be revealed through a deconstructionist strategy – manifests at two levels:

1) an external power exercised over cartography by a third entity seeking to impose particular norms and value systems on the greatest number of people (typically the State and the Church);

2) an internal power that originates in the very process of map-making (selections and omissions from the cartographer) and which is generally equated with cartographic intentionality (Noucher 2015).

15 On September 3, 2020, Google wrote a post about this topic on its official blog ("Google Maps 101: How AI helps predict traffic and determine routes").

While the position defended by Harley remains debatable in certain respects – especially from the standpoint of his understanding and rendering of Foucault's and Derrida's works (see Belyea 1992; Crampton 2001) – his deconstructionist practice of the map remains useful for elucidating the mechanisms underlying the production of the geographic *data/capta/quaesita* triptych.

2.4.1. *Geographic data: striving for ever greater accuracy*

This subsection offers an overview of the deconstruction of geographic *data* published on OSM. Unlike *data* created and distributed by public institutions like the French National Institute of Cartography (IGN) or private companies like Google, these are (supposedly) produced by local actors who know the "reality on the ground". For this reason, OSM is often presented as the Web 2.0 platform that opened up and democratized cartographic practices that were once reserved for experts. Encouraged to follow a set of best practices[16], OSM users are free to feed the geographic database (nodes, paths and relations associated with semantic information commonly called tags) and the platform's discussion threads. This bottom-up approach contrasts with conventional cartographic production, which may lead more than one person to think – wrongly – that OSM data escape the implicit power(s) highlighted by Harley. As early as the 2010s, Elwood (2009, 2010) invited the scientific community to examine the socio–technical–political factors that were (and still are) at play in the production of geographic data on the web. Work conducted over the past fifteen years shows that these factors govern the fabric of geographic *data* on the web, including those that result from a participatory process.

To start with, like the over-representation of Western European places on Wikipedia (Dittus and Graham 2019), we observe a gap in terms of geographic coverage on OSM, although it has been decreasing over the past five years. Despite the proliferation of mapping parties (e.g. Mapathons), OSM *data* remain far more comprehensive in European and North American countries (Schröder-Bergen et al. 2021). It also appears that most of the contributors are men who have technical skills and live in the northern hemisphere (Glasze and Perkins 2015), which is consistent with the facts advanced by Crampton et al. (2013), namely that the collection of user-generated data is generally carried out by "more wealthy, more educated, more Western, more white and more male demographic" (p. 132). Moreover, OSM data covering developed countries are mostly produced by local actors, as well as, more

16 For more information, see: https://wiki.openstreetmap.org/wiki/Good_practice.

surprisingly, by private companies[17] (Anderson et al. 2019). Particularly active since 2017, these have an interest in contributing to the collaborative mapping platform's database. Amazon's delivery service, for example, uses OSM data: based on feedback from its delivery drivers, the company continuously enriches road network data, mainly on the American continent and in Europe. Meta, for its part, favors automated contributions supervised using machine learning algorithms. Other digital giants, such as Microsoft and Uber, also participate in creating and updating OSM's geographic *data*. Apple and MapBox stand out from other private actors by operating on the six continents.

At the same time, data production on OSM in Least Developed Countries and Developing Countries is carried out by international humanitarian organizations and local actors (Schröder-Bergen et al. 2021). In this context, the mapping of spaces classified as "sensitive" (conflict zones, areas devastated by climate hazards, etc.) has also highlighted socio–technical–political factors veiled by the alleged objectivity of OSM *data*. The Israeli–Palestinian mapping of Jerusalem deserves to be cited as an example. Precisely, Bittner (2017) showed that the density of OSM data was significantly higher in neighborhoods occupied by Israeli Jews (469 nodes, 64 paths and 182 tags per 1,000 people) compared to those inhabited by Palestinian Arabs (138 nodes, 15 paths and 56 tags per 1,000 people). This disparity is logically found in the languages used in entity labels: 1.5% of OSM tags are written in Arabic letters versus 12.8% in Hebrew; this ratio is rising to 1.2/14% for paths. Also, it appears that only 15 users out of 163 are responsible for 95% of contributions on Jerusalem. Several difficult-to-identify factors can explain these disparities (Internet access, technical skills, external influences, etc.).

Interestingly, Bittner (2017) observes the exact opposite on Wikimapia[18]. In Jerusalem, the alternative collaborative mapping platform is indeed far more invested in by the Arab community. Bittner and Glasze explain this phenomenon by an epistemological divergence:

> While OSM pursues an agenda of creating a geometrically accurate representation of the world according to facts on the ground,

17 See, in particular, Mikel Maron's presentation at the 2020 edition of the State of the Map international conference: https://2020.stateofthemap.org/sessions/RHDUV9/, and read in Chapter 3 about how these major web companies are involved in OSM and the Overture Maps initiative, which signals a change in this area.

18 Wikimapia is a collaborative online mapping platform. Launched in 2006, it allows users to mark and describe places on a world map. Unlike OSM, which focuses on creating a free and comprehensive geographic database, Wikimapia focuses on annotating places by encouraging its users to add descriptions, photos and links for specific points of interest.

> Wikimapia rather seeks to collect semantic descriptions of places. We argue that OSM's approach inherently tends to reproduce spatial orders and leaves the power imbalances that underly these orders unquestioned, which results in a map that is hard to accept and probably unworthy of support from a Palestinian point of view. Wikimapia's agenda, however, is generally more open for confrontational, playful, or even artistic forms of mapping. (Bittner and Glasze 2021, pp. 64–65)

Indeed, the production of OSM data is dictated by the "Truth on the ground" principle according to which they must match the reality on the ground, both from a material and sociocultural point of view. Thus, because they are intended to be represented on a conventional map, these geographic *data* tend ipso facto to impose a certain form of "ground truth" that, in fact, is contested through edit wars[19] and digital vandalism (Ballatore 2014; Truong et al. 2018). Unsurprisingly, algorithmic solutions to improve the quality of these data essentially target their geometric, semantic and temporal accuracy (Girres and Touya 2010).

2.4.2. *The illusion of geographic capta*

As part of Big Data, geographic *capta* inherit de facto their attributes: they would thus be exhaustive geographic "data" (i.e. concerning a population and no longer a sample) generated in abundance and in real time by citizens themselves (Kitchin 2014; Quesnot 2016a). Their supposed characteristics suggest that they offer privileged access to individual and collective place practices. While the abundance of geographic *capta* is certainly undeniable, their exhaustive nature and their real-time production by social media users are, however, debatable. The development of Web 2.0, combined with the proliferation of smartphones and the rise of mobile Internet and geolocation, has indeed precipitated this deluge of geographic data. Today, the vast majority, if not all, social web users participate voluntarily (or not) in this saturation of geographic *capta* on the web.

Overall, it is important to emphasize that *capta* are not representative of the general population. It is estimated, for example, that Twitter (X) represents one-third of Internet users and that the Twittersphere contains only between 1% and 2% of geotagged tweets (Ajao et al. 2015; Karami et al. 2021). Also, all digital social media (Facebook, Instagram, X, etc.) that allow access to their data – free or paid – generally proceed with opaque filtering that is difficult, if not impossible, to

19 For more information, see: https://wiki.openstreetmap.org/wiki/Disputes.

circumvent, as the algorithms are never disclosed (Rzeszewski 2016). Moreover, these platforms have a difficult-to-quantify number of bots that carry out semi-automated posts[20]. Ferrara et al. (2016) define these social media bots as "computer algorithm that automatically produces content and interacts with humans on social media, trying to emulate and possibly alter their behavior" (p. 96). Several families of bots operate on the social web. These can be neutral, like those that automatically post news, benevolent (typically, chatbots) or malicious (e.g. spam bots or Sybils).

With the power and widespread use of generative AIs (ChatGPT, Claude, Gemini and more recently DeepSeek), the impact of these algorithms is gigantic, both at the individual and collective scale. Astroturfing bots, for example, are capable of influencing public opinion by creating support for (or conversely opposition to) an individual (e.g. a presidential candidate) from scratch using comments, likes and other similar content. Others specialize in political conflicts. These seek to divert readers' attention by polluting discussion threads through false or completely off-topic assertions (for a complete overview, see Orabi et al. 2020). Recently, Himelein-Wachowiak et al. (2021) estimated that bots were behind 66% of discussions on Twitter around Covid-19 and that they actively participated in disinformation on this subject. Ultimately, the rise of these algorithms and their significant impacts call for caution: conclusions drawn from the analysis of (geographic) *capta* must be approached with circumspection. In reality, we have no information about the context in which social web users generated them. This acontextuality, being characteristic of *capta*, should be kept in mind that the media from which they are extracted remain above all a means of communication, and that motivations for use, which are de facto multiple (social interactions, entertainment, information seeking, evaluation of a place, ego valorization, etc.), vary greatly depending on the platform selected (Oh and Syn 2015).

Despite this, some projects do not hesitate to mix geographic *capta* together to increase their claimed potential. This is the case, for example, with the Social Media Ingestor (Stefanidis et al. 2013), which integrates data from multiple social web platforms such as Twitter, Flickr and YouTube, with the aim of analyzing social networks and information flows in a geographic context, without however resolving the problems inherent to each data source or questioning the validity of this heterogeneous aggregation. Beyond the fact that these practices do not take into account the motivational incompatibilities mentioned above, they reveal a singular phenomenon that we call "positional primacy" (Quesnot and Roche 2015b; Quesnot 2016a), which consists of making position – modeled here in the form of a

20 It is worth remembering that behind these bots, there are inevitably one or more individuals.

geolocation tag (geotag) – the privileged receptacle of the ideational dimension of geographic space. This reduction of place to a geotag augmented with allegedly user-generated geographic information participates in the movement of (over-)quantification of human spatialities. It reinforces the Cartesian conception of space according to which it is assimilated to a simple support for interactions. Also, many researchers maintain that geographic *capta* allow for concrete apprehension of the concept of place and the multiple meanings we attach to it (Jenkins et al. 2016). This line of reasoning leads to a fallacious shortcut: assuming that positional equivalence ipso facto translates into ideational equivalence. It suggests, for example, that a geotagged tweet and a geolocated Instagram photo published from the same place necessarily deal with that place, when this is not necessarily the case[21].

Why do social media users geolocate their content? Is it a conscious act, or a practice implicitly encouraged, or even forced, by the platform? How does the geographic information uploaded by these users change our way of perceiving the immediate environment? Only qualitative research (interviews, observations, focus groups, questionnaires, etc.) can answer these questions raised, among others, by Rzeszewski (2016).

2.4.3. *Geographic quaesita and algorithmic spatialities*

At one point or another, all (geo)digital footprints on the web go through one or even several algorithmic filters. This transformation necessarily involves a black box phenomenon that a part of the scientific community invites us to decode[22] (Noucher 2023). In fact, the opacity of the treatments applied depends on the footprint under consideration. Algorithms related to geographic *data* mainly concern the improvement of their geometric accuracy (location and topology) and their geographic coverage. When these *data* are intended to be mapped, like those of OSM and Wikimapia, algorithms dealing with their formatting (symbols, labels and membership categories) add an additional layer to decipher. Regarding these two platforms in particular, open-source code and discussion threads (Wiki) are of invaluable help to anyone wishing to deconstruct the *data* they shape and host. We move up a notch in terms of opacity with the production and dissemination of geographic *capta*. Several algorithmic layers intercede between their production by

21 Indeed, as another example, a tweet geolocated near the Eiffel Tower may refer to a place different from the Parisian monument.

22 Another part of our community argues that the pursuit of algorithmic transparency and explainability at all costs is not necessarily desirable (Anany and Crawford 2016; Zerilli et al. 2019).

users and their reception by a third party (aggregations, filtering operated by the API, etc.). Since these algorithms are proprietary, it is therefore impossible for us to access them. At the moment, the only deconstructive approach that seems viable is that of inquiry into the perceptions and actual uses of these platforms.

Deconstructing geographic *quaesita* is also problematic, since they are manufactured then dissolved in "real time" by closed algorithms. Would their deconstruction therefore be utopian? While this is certainly a delicate undertaking, a deconstructionist strategy does appear possible, provided that attention is focused on the appropriate target, namely the spatiality of *quaesita*, rather than the algorithm that creates them. More specifically, the arrival of geographic *quaesita* in the sphere of the (geo)web is very recent and largely dependent on the democratization of augmented reality (AR) and its opposite, diminished reality[23] (DR) (Mori et al. 2017), whose development should intensify with the recent market release of AR/DR/mixed reality (MR)[24] headsets and glasses. Under these conditions, the discussion presented here can therefore only be prospective. It is based on current trends and has no other ambition than to provide a framework for understanding this emerging and algorithmic practice, which is egography (Quesnot 2021), and to propose a first avenue for deciphering the behavior of algorithms that shape the egographic filters of AR/DR/MR.

Geographic *quaesita* are already distilled on popular mapping platforms like Google Maps and CityMapper, and their rise could prove considerable with the imminent democratization of reality technologies and spatial computing. In short, AR/DR/MR enables the emergence of a representation of geographic information complementary to cartography: egography, which we define as "the practice of appropriately arranging and representing these elements (AR/DR/MR filters) to orient movement for a specific purpose (navigation assistance, search for a place, valorization of a shop, estimation of time-distance, etc.)" (Quesnot 2021). It is therefore not AR/DR/MR per se, but nevertheless exploits this technology to display annotations and digital pictograms in real time, and also to obscure or even mask elements of the immediate environment. The conventional map projects onto a

23 Unlike augmented reality, which aims to add layers of information to the immediate physical environment, diminished reality involves removing certain parts of reality (e.g. blurred backgrounds on Zoom).

24 Mixed reality combines augmented reality and virtual reality. It offers the ability to manipulate a virtual space by integrating elements from the physical world. This technology requires the use of a special headset equipped with a screen that allows the user to simultaneously view the real and virtual worlds, with the addition of layers of information that integrate "naturally" into the user's field of perception.

medium with restricted dimensions (typically a sheet of paper or the screen of a digital device) a reduced and simplified representation of a portion of the physical environment, whereas egography uses the latter in extenso (1:1 scale) to show information, images and symbols. Unlike cartography, which is often approached from the angle of cognitivism[25], egography is experiential by essence, because it requires the engagement of the body. Therefore, it participates in reinforcing the paradigm of situated and embodied cognition (Varela et al. 1993).

Currently, the AR/DR/MR market is booming. Snapchat Inc. has become a major player in this field with the popularization of its ego filters. It continues to invest in the AR/DR domain and now targets the advertising sector[26]. Digital giants like Apple and Meta are also in the game with the development of their mixed reality headsets (AR and VR) (respectively Apple Vision Pro and Meta Quest). Google is not left behind. Like Apple and its famous ARKit, the company released its ARCore kit in 2018, dedicated to the development of AR/DR/MR applications. It has been developing a new communication solution for several years, allowing dialogue through interposed holograms (Project Starline). Today, Google offers three AR services for mobile devices:

1) Google Lens is used to identify objects and entities (plants, animals, etc.) and translate text.

2) Its search engine now includes a feature that allows displaying search results in 3D and AR.

3) Google Maps has a Live View function that allows pedestrians to benefit from AR navigation assistance.

All of these services are currently available on smartphones and tablets, but the imminent return of AR glasses to the civilian sphere (see Meta's Project Orion) will contribute to the democratization and therefore massive use of this technology, which necessarily raises a number of issues, particularly ethical ones.

Indeed, while Google Maps' AR functionality currently only contextualizes a user's movement (i.e. displays in real time "information" already viewable on the classic version of the mapping platform), it is likely that geographic *quaesita* will come to orient users' future movements. Since Google derives the vast majority of

25 The cognitivist approach equates cognition with a mind that interprets the objects (in the Cartesian sense of the term) surrounding it in the form of symbolic representations.

26 At the end of 2021, Snapchat Inc. partnered with the giant WPP to produce AR marketing solutions.

its profits from advertising, chances are that these *quaesita* will be partially displayed to promote a place in various ways (unique symbol that contrasts with the rest, notification, blurring or even disappearance of elements through DR, etc.). Beyond this marketing aspect, it appears essential to study in depth how these AR/DR/MR filters will shape our representations of space and therefore our relationship to the world. And for good reason: experimental psychology research has demonstrated that our memory is intrinsically linked to our way of dividing the spaces we perceive in one go. For example, crossing what we consider to be a threshold (like a door) tends to cause a loss of short-term memory (objects, words, images, etc.). This phenomenon is verified in material environments (Radvansky et al. 2011), virtual ones (Radvansky and Copeland 2006) and even imaginary ones (Lawrence and Peterson 2016). Since AR/DR/MR filters and the geographic *quaesita* that compose them can potentially generate and accentuate these thresholds, careful examination of their real-time manifestation on these technologies proves essential.

Ultimately, it is not the content of the algorithms as such that we must attempt to decipher – this enterprise being futile, since we are dealing with proprietary algorithms – but the mode of representation of geographic *quaesita*, as well as their positioning "on" the physical environment. In other words, deconstructing them amounts to nothing more than analyzing their spatialities and the ways in which they appear (rather than are) in the world. From an experimental perspective, these two parameters – positioning and displaying of geographic *quaesita* – would serve as dependent variables, while we – users of AR/DR/MR technologies – could assume the role of independent variables. Since the algorithms that deploy AR/DR/MR filters are a function of our movements through the environment and controllable individual parameters (browsing history, gender, etc.), it would then be possible to observe to what extent our own spatialities impact those of geographic *quaesita*[27].

Under these conditions, future research in the field of egography would be well advised to focus on the following questions to anticipate and prepare society for these potential technological developments: In what form(s) do AR/DR/MR filters present themselves to the user? Does this way of appearing vary according to individual factors? Do we observe the same geographic *quaesita* for the same route but a different user? In which case, what are the factors that determine the appearance of a *quaesita*? In addition, it would be interesting to find out how these

27 For example, an artist named Simon Weckert fooled Google Maps' traffic prediction by dragging smartphones behind him: https://www.youtube.com/watch?v=k5eL_al_m7Q&t=1s (accessed December 12, 2024).

AR/DR/MR filters influence our own daily behaviors. The focus would then shift from geographical *quaesita* to the users of these technologies.

2.5. Conclusion

For nearly half a century, the creation of geographic data was exclusively reserved for the government and a few private companies working mainly in the military sector. The arrival of Web 2.0 and collaborative mapping platforms in the early 2000s completely changed the game. On the one hand, citizens became major producers of geographic data, now considered potentially reliable by actors in both public and private spheres. On the other hand, the participatory dimension that characterizes this second generation of the web led to the development of digital social media from which Internet users publish a huge volume of digital content. Given the diversity of these publications, it was essential to establish a typology of geographic "data" on the web and to propose an initial deconstructive reading of them. In summary, geographic *data* published on the web refer to all data that deal with the material component of the physical environment. Even though they are artifacts, these data are those that come closest to the original meaning of the word *data* (i.e. a given thing or quantity). That said, by virtue of their vocation to represent the material reality on the ground as faithfully as possible, geographic *data* are subject to the same issues raised by Harley (1989) regarding the map. *Capta* that emanate from the social web differ from geographic *data*. These are geolocated content – and not necessarily by primary intention – whose primary vocation is communication. As the etymology of the word *capta* suggests, these digital publications are seized by social web platforms, transformed and then redistributed. Contrary to appearances, they are not a neutral reflection of our affective and symbolic relationships to the environment. *Quaesita*, meanwhile, emerged with the rise of geolocation and the geospatial footprints it enables. Experiential per se, the geographic *quaesita* disseminated on popular mapping platforms not only contextualize our movements: they orient them for a specific purpose by subtly inducing behavioral changes.

With the democratization of augmented/diminished/mixed reality (AR/DR/MR) technologies, it is highly likely that geographic *quaesita* will be increasingly mobilized in the near future. The study of their egographies – that is, the way they overlay our physical environment through AR/DR/MR filters – therefore deserves to be seriously addressed by the scientific community.

2.6. References

Ajao, O., Hong, J., Liu, W. (2015). A survey of location inference techniques on Twitter. *Journal of Information Science*, 41(6), 855–864.

Ananny, M. and Crawford, K. (2016). Seeing without knowing: Limitations of the transparency ideal and its application to algorithmic accountability. *New Media & Society*, 20(3), 973–989.

Anderson, J., Sarkar, D., Palen, L. (2019). Corporate editors in the evolving landscape of OpenStreetMap. *ISPRS International Journal of Geo-Information*, 8(5). doi: 10.3390/ijgi8050232.

Aschan-Leygonie, C., Cunty, C., Davoine, P.-A. (2019). *Les systèmes d'information géographique. Principes, concepts et méthodes*. Armand Colin, Paris.

Ballatore, A. (2014). Defacing the map: Cartographic vandalism in the digital commons. *The Cartographic Journal*, 51(3), 214–224.

Belyea, B. (1992). Images of power: Derrida/Foucault/Harley. *Cartographica: The International Journal for Geographic Information and Geovisualization*, 29(2), 1–9.

Bittner, C. (2017). Diversity in volunteered geographic information: Comparing OpenStreetMap and Wikimapia in Jerusalem. *GeoJournal*, 82(5), 887–906.

Bittner, C. and Glasze, G. (2021). Who maps Middle Eastern geographies in the digital age? Inequalities in web 2.0 cartographies in Israel/Palestine. In *Media and Mapping Practices in the Middle East and North Africa*, Strohmaier, E. and Krewani, A. (eds). Amsterdam University Press, Amsterdam.

Chrisman, N.R. (2004). *Charting the Unknown: How Computer Mapping at Harvard Became GIS*. ESRI Press, Redlands.

Crampton, J. (2001). Maps as social constructions: Power, communication and visualization. *Progress in Human Geography*, 25(2), 235–252.

Crampton, J., Graham, M., Poorthuis, A., Shelton, T., Stephens, M., Wilson, M.W., Zook, A.M. (2013). Beyond the geotag: Situating 'big data' and leveraging the potential of the geoweb. *Cartography and Geographic Information Science*, 40(2), 130–139.

Cranshaw, J., Schwartz, R., Hong, J.I., Sadeh, N. (2012). The livehoods project: Utilizing social media to understand the dynamics of a city. In *Proceedings of the Sixth International AAAI Conference on Weblogs and Social Media*. AAAI Press, Washington, D.C.

Daston, L. and Galison, P. (2012). *Objectivité*. Les presses du réel, Dijon.

Dittus, M. and Graham, M. (2019). Mapping Wikipedia's geolinguistic contours. *Digital Culture & Society*, 5(1), 147–164.

Drucker, J. (2011). Humanities approaches to graphical display. *Digital Humanities Quarterly*, 5(1), 1–21.

Elwood, S. (2009). Geographic information science: Emerging research on the societal implications of the geospatial web. *Progress in Human Geography*, 34(3), 349–357.

Elwood, S. (2010). Geographic information science: Visualization, visual methods, and the geoweb. *Progress in Human Geography*, 35(3), 401–408.

Ferrara, E., Varol, O., Davis, C., Menczer, F., Flammini, A. (2016). The rise of social bots. *Communications of the ACM*, 59(7), 96–104.

Fischer, F. (2012). VGI as big data. A new but delicate geographic data-source. *Geoinformatics*, 15(3), 46–47.

Galinon-Mélénec, B. (2013a). Expérience incarnée, construction cognitive et jugement. *Revue française des Sciences de l'Information et de la Communication*. 3. doi: 10.4000/rfsic.487.

Galinon-Mélénec, B. (2013b). Des signes-traces à l'Homme-trace. La production et l'interprétation des traces placées dans une perspective anthropologique. *Intellectica. Revue de l'Association pour la Recherche Cognitive*, 59(1), 89–113.

Galinon-Mélénec, B. (2015). À la recherche de la trace. *Communication et organisation*, 47, 31–50.

Garson, G.D. and Biggs, R.S. (1992). *Analytic Mapping and Geographic Databases (Issue 87)*. SAGE Publications, Newbury Park.

Girres, J.-F. and Touya, G. (2010). Quality assessment of the French OpenStreetMap dataset. *Transactions in GIS*, 14(4), 435–459.

Glasze, G. and Perkins, C. (2015). Social and political dimensions of the OpenStreetMap project: Towards a critical geographical research agenda. In *OpenStreetMap in GIScience. Experiences, Research, and Applications*, Arsanjani, J.J., Zipf, A., Mooney, P., Helbich, M. (eds). Springer, Cham.

Goodchild, M.F. (2007). Citizens as sensors: The world of volunteered geography. *GeoJournal*, 69(4), 211–221.

Hahmann, S. and Burghardt, D. (2013). How much information is geospatially referenced? Networks and cognition. *International Journal of Geographic Information Science*, 27(6), 1171–1189.

Haklay, M. and Weber, P. (2008). OpenStreetMap: User-generated street maps. *IEEE Pervasive Computing*, 7(4), 12–18.

Harley, J.B. (1989). Deconstructing the map. *Cartographica: The International Journal for Geographic Information and Geovisualization*, 26(2), 1–20.

Himelein-Wachowiak, M., Giorgi, S., Devoto, A., Rahman, M., Ungar, L., Schwartz, H.A., Epstein, D.H., Leggio, L., Curtis, B. (2021). Bots and misinformation spread on social media: Implications for COVID-19. *Journal of Medical Internet Research*, 23(5). doi: 10.2196/26933.

Jenkins, A., Croitoru, A., Crooks, A.T., Stefanidis, A. (2016). Crowdsourcing a collective sense of place. *PLoS ONE*, 11(4). doi: 10.1371/journal.pone.0152932.

Karami, A., Kadari, R.R., Panati, L., Nooli, S.P., Bheemreddy, H., Bozorgi, P. (2021). Analysis of geotagging behavior: Do geotagged users represent the Twitter population? *ISPRS International Journal of Geo-Information*, 10(6). doi: 10.3390/ijgi10060373.

Kitchin, R. (2014). Big data, new epistemologies and paradigm shifts. *Big Data & Society*, 1(1), 1–12.

Lavin, M. (2021). Why digital humanists should emphasize situated data over capta. *Digital Humanities Quarterly*, 15(2) [Online]. Available at: https://digitalcommons.denison.edu/cgi/viewcontent.cgi?article=2066&context=facultypubs.

Lawrence, Z. and Peterson, D. (2016). Mentally walking through doorways causes forgetting: The location updating effect and imagination. *Memory*, 24(1), 12–20.

Lindqvist, J., Cranshaw, J., Wiese, J., Hong, J.I., Zimmerman, J. (2011). I'm the mayor of my house: Examining why people use foursquare – A social-driven location sharing application. In *Proceedings of the Conference on Human Factors in Computing Systems (CHI-11)*. ACM, Vancouver.

Longino, H.E. (1990). *Science as Social Knowledge*. Princeton University Press, Princeton.

Lussault, M. (2007). *L'homme spatial : la construction sociale de l'espace humain*. Le Seuil, Paris.

Mericskay, B., Noucher, M., Roche, S. (2018). Usages des traces numériques en géographie : potentiels heuristiques et enjeux de recherche. *L'Information Géographique*, 82(2), 39–61.

Mori, S., Ikeda, S., Saito, H. (2017). A survey of diminished reality: Techniques for visually concealing, eliminating, and seeing through real objects. *IPSJ Transactions on Computer Vision and Applications*, 9. doi: 10.1186/s41074-017-0028-1.

Noucher, M. (2015). De la trace à la carte et de la carte à la trace : pour une approche critique des nouvelles sources de fabrique cartographique. In *Traces numériques et territoires*, Severo, M. and Romele, A. (eds). Presses des Mines, Paris.

Noucher, M. (2023). *Blancs des cartes et boîtes noires algorithmiques*. CNRS Éditions, Paris.

O'Reilly, T. (2005). What is web 2.0: Design patterns and business models for the next generation of software [Online]. Available at: http://oreilly.com/web2/archive/what-is-web-20.html [Accessed July 4, 2024].

Oh, S. and Syn, S.Y. (2015). Motivations for sharing information and social support in social media: A comparative analysis of Facebook, Twitter, Delicious, YouTube, and Flickr. *Journal of the Association for Information Science and Technology*, 66(10), 2045–2060.

Orabi, M., Mouheb, D., Al Aghbari, Z., Kamel, I. (2020). Detection of bots in social media: A systematic review. *Information Processing & Management*, 57(4). doi: 10.1016/j.ipm.2020.102250.

Ostrom, E. (1990). *Governing the Commons*. Cambridge University Press, Cambridge.

Quesnot, T. (2016a). L'involution géographique : des données géosociales aux algorithmes. *NETCOM*, 30(3–4), 281–304.

Quesnot, T. (2016b). Linked landmark data: Toward the detection of landmarks on the web of data. In *Advancing Geographic Information Science: The Past and the Next Twenty Years*, Onsrud, H. and Kuhn, W. (eds). GSDI Association Press, Needham, MA.

Quesnot, T. (2021). Décrypter l'innovation de la cartographie numérique grand public : entre marchandisation, appropriation et représentation égographique de l'information. *Mappemonde*. 131. doi: 10.4000/mappemonde.5728.

Quesnot, T. and Roche, S. (2015a). Platial or locational data? Toward the characterization of social location sharing. In *Proceedings of the 48th Annual Hawaii International Conference on System Sciences*, Bui, T.X., Ralph, H., Sprague, J. (eds). IEEE Computer Society Press, Los Alamitos, CA.

Quesnot, T. and Roche, S. (2015b). Measure of landmark semantic salience through geosocial data streams. *ISPRS International Journal of Geo-Information*, 4(1), 1–31.

Quesnot, T. and Roche, S. (2020). Explorer les mobilités partagées à l'ère des technologies de la géolocalisation. In *Médias situés et mobilités partagées*, Guelton, B. (ed.). Hermann, Paris.

Radvansky, G.A. and Copeland, D.E. (2006). Walking through doorways causes forgetting: Situation models and experienced space. *Memory & Cognition*, 34(5), 1150–1156.

Radvansky, G.A., Krawietz, S.A., Tamplin, A.K. (2011). Walking through doorways causes forgetting: Further explorations. *Quarterly Journal of Experimental Psychology*, 64(8), 1632–1645.

Rost, M., Barkhuus, L., Cramer, H., Brown, B. (2013). Representation and communication: Challenges in interpreting large social media datasets. In *Proceedings of the 2013 Conference on Computer Supported Cooperative Work*, Bruckman, A. and Counts, S. (eds). ACM, New York.

Rouvroy, A. (2017a). La vie n'est pas donnée. *Études digitales*, 2016(2), 195–217.

Rouvroy, A. (2017b). Penser le sujet comme puissance : contingence et potentialité à l'ère du calcul intensif. In *Ni objet ni robot : simplement humain*. Grappe, Bois-de-Villers.

Rouvroy, A. and Berns, T. (2013). Gouvernementalité algorithmique et perspectives d'émancipation : le disparate comme condition d'individuation par la relation ? *Réseaux*, 177, 163–196.

Rzeszewski, M. (2016). Geosocial capta in geographical research – A critical analysis. *Cartography and Geographic Information Science*, 45(1), 18–30.

Schröder-Bergen, S., Glasze, G., Michel, B., Dammann, F. (2021). De/colonizing OpenStreetMap? Local mappers, humanitarian and commercial actors and the changing modes of collaborative mapping. *GeoJournal*, 87, 5051–5066.

Stefanidis, A., Crooks, A., Radzikowski, J. (2013). Harvesting ambient geospatial information from social media feeds. *GeoJournal*, 78(2), 319–338.

Truong, Q.T., Touya, G., De Runz, C. (2018). Le vandalisme dans l'information géographique volontaire : apprendre pour mieux détecter ? In *Actes de la conference SAGEO*, Montpellier.

Varela, F., Thompson, E., Rosch, E. (1993). *L'inscription corporelle de l'esprit : sciences cognitives et expérience humaine*. Le Seuil, Paris.

Zerilli, J., Knott, A., Maclaurin, J., Gavaghan, C. (2019). Transparency in algorithmic and human decision-making: Is there a double standard? *Philosophy & Technology*, 32(4), 661–683.

3

From a Database to a Digital Commons: The OpenStreetMap Project

Boris MERICSKAY
UMR Espaces et Sociétés (ESO–Rennes), Université Rennes, France

3.1. Introduction

Created in 2004 in the United Kingdom, the OpenStreetMap (OSM) project has profoundly changed the landscape and uses of geospatial data. Similar to Wikipedia, this collaborative project, which aims to create an open vector geographic database, has become a key player in the global mapping ecosystem in just a few years. From an initiative led by a handful of enthusiasts, OSM has become increasingly professional over the years and is now a source of spatial data used by a wide range of users. Beyond being a simple free and open spatial database, OSM, through its multifaceted and multi-stakeholder nature, can now be considered a digital commons. It serves as a driver of cooperation, innovation and new forms of interaction between the public sector, the private sector and civil society.

With more than 11 billion geographic objects in its database and a community of over 10 million contributors, OSM is now one of the best examples of what can be achieved when digital mapping and collaborative online practices come together. By combining open data, voluntary contributions and free software development, the project positions itself as a free alternative to the privatization and centralization of

Geospatial Technologies and Society,
coordinated by Thierry JOLIVEAU and Matthieu NOUCHER. © ISTE Ltd 2026.

spatial data promoted by digital giants. From simple individual contributions to state production programs, field data collection campaigns and the development of dedicated applications, OSM brings together thousands of participatory mapping initiatives with diverse themes and objectives. The project's specificity and strength lie in the volume and diversity of its contributors. Its organization is based on a decentralized system that brings together national, local, technical and thematic communities, as well as computer scientists, cartographers, activists, academics and private companies.

Since its foundation 20 years ago, OSM has largely moved beyond the realm of amateurism. Contributors have become increasingly specialized and organized; the infrastructure and associated services are growing more technically advanced, and the data produced continue to expand in both volume and quality. At the same time, the growing involvement of large companies such as Microsoft, Meta and Amazon is transforming both the development and governance of the project. To understand the functioning, role and status of geographic data in the digital and Internet age, it is important to consider how and why OSM is coming to occupy an increasingly significant place in the world of digital cartography, and what challenges are associated with its development.

The aim of this chapter is to provide a broad overview of the OSM project, covering its technical foundations, evolution, community, associated practices and prospects in terms of geographic data and stakeholder groups. By presenting the diversity of initiatives, modes of governance, issues of institutionalization, and coordination with public and private actors, we aim to explore the mechanics of this ecosystem and highlight its various facets. This chapter examines OSM from a sociotechnical perspective, explaining how this original digital commons project is renewing and transforming the ways in which geographic data are produced and used.

The first part provides a brief history of the project and explains its technical workings. The second part examines the OSM community, a diverse group organized and structured around projects and territories. The question of interactions between OSM and the operational world is then addressed through an overview of the different types of partnerships with public and private actors. Finally, several prospective issues related to the project are presented and discussed, particularly those concerning data quality, the development of 3D modeling, indoor mapping and the future of the project in the era of digital commons.

3.2. The OSM project: its history and how it works

3.2.1. *A brief history of the OSM project*

The OSM project was founded in 2004 by Steve Coast, then a computer science student at University College London. His main motivation was based on the fact that the British mapping agency, Ordnance Survey, retained the copyright on geographical data for its own benefit, even though it was funded by taxpayers (Coast 2015). In response to this situation, he came up with the idea of setting up a collaborative geographic database under a free license, which could be edited and used by anyone without financial compensation.

From 2004 to 2006, the project remained fairly low-key, with a few thousand contributors feeding the database with GPS tracks collected by bicycle to digitize roads. In 2006, Yahoo imagery was integrated into the online editor, changing the way data were produced. Instead of starting with a blank map and going out into the field to collect data, it was now possible to use images to digitize elements around the world from behind a screen. Given the project's success, the OSM Foundation (OSMF) was created in August 2006 to manage the necessary hardware infrastructure and provide legal protection for the data and contributors.

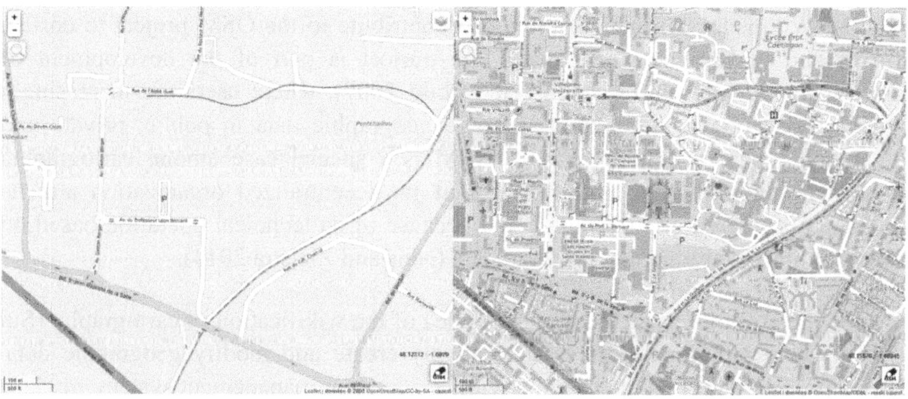

Figure 3.1. *Changes in OSM coverage around the Rennes 2 University campus between 2008 and 2024 (source: OSM). For a color version of this figure, see www.iste.co.uk/joliveau/geovisualization.zip*

From 2007 to 2009, the project began to expand in terms of the number of contributors, their organization and data production. Several private companies and public bodies began to provide databases free of charge to feed and enrich the project. For example, in 2007, the Dutch company Automotive Navigation Data offered a complete database of the Dutch road network. The TIGER database published by the U.S. Census Bureau (road network, administrative divisions and hydrography) was fully integrated into the OSM map that same year.

In 2010, the detailed mapping of Haiti produced by the community in 48 h following the earthquake enabled the project to gain international recognition and visibility (Palen et al. 2015). That same year, the integration of aerial imagery from the giant Bing (Microsoft) into a new, more user-friendly and powerful editor further simplified data production methods. With more than 100,000 contributors by the end of 2010, the project began to take on a more structured international dimension with the organization of annual meetings (States of the Map), the development of local branches around the world and the transition to the Open Database License. The milestone of 1 million contributors was reached in 2013, rising to 2 million in 2015, 5 million in 2017 and 8 million in 2021. Since then, the project has continued to grow in terms of contributors, data produced, related initiatives and technical instrumentation.

3.2.2. *A collaborative and open mapping project*

Like Wikipedia, all internet users can contribute to the OSM project to enrich, correct and update the world map. The project is part of the development of voluntary geographic information (Goodchild 2007), where users are increasingly being called upon to produce and update geographic data in public, private and community contexts (Heipke 2010). OSM is a special case among cartographic crowdsourcing platforms, both because of its decentralized organization and its philosophy centered on “freedom”, and because of its technical operation based on the collaborative production of spatial data (Neis and Zielstra 2014).

OSM relates more specifically to the idea of the wikification of cartography (Sui 2008), where individuals can collaboratively create and modify geographic data. This approach, based on the use of a wiki as a content management system, marks a technological and methodological break in the world of GIS (Roche et al. 2012). Wikis are an example of “social software” that allows for the iterative production of content while offering a set of tools for contributors to organize and communicate with each other. Applied to the production of geographic data, they offer the possibility of going beyond the individual use of GIS and promote the dynamic and collective construction of geographic data through the exchange of different knowledge and points of view.

The OSM database consists of geographical objects found in public spaces. Only elements that are visible and clearly defined on the ground (buildings, road networks, street furniture, shops, trees, etc.) or formally recognized (administrative boundaries, bus routes, etc.) can be mapped. This convention ensures that other contributors can compare the real object with its model in the database to modify or update it if necessary.

There are three main ways to contribute to the OSM database: by adding GPS tracks collected in the field, by digitizing features via aerial or satellite images or by incorporating spatial data produced by third parties (public or private) with a compatible license.

All data in the OSM database is available under the free Open Database License (ODbL). This license is designed for databases and clearly specifies the conditions for reuse. Anyone is free to copy, distribute, adapt or even monetize the data, provided they credit OSM and its contributors and make the new data available under the same license. The data are accessible via a multitude of services that allow it to be viewed as base maps or retrieved via multiple channels (features within GIS software, download platforms, geographic web services and application programming interfaces).

3.2.3. *Mapping the world in OSM*

In traditional GIS vector data, the world features are modeled as points, lines and polygons, which are associated with attribute tables. In the OSM database, geometric objects are represented through three fundamental elements: nodes, ways and relations. (Figure 3.2). From a geometric point of view, all cartographic elements are based on nodes. Consisting of a pair of geographic coordinates (in the WGS84 coordinate system), these basic elements can be used alone to model an address, a school, a business or a tree.

When connected together, nodes can be used to model linear or zonal objects. These are referred to as ways, which can be open to map roads, rivers, bike paths, etc., or closed to represent buildings, parks, forests, lakes, etc. In addition, relations are used to group different objects together, which, when considered as a whole, form a new object. Examples include different sections of road forming a bus route, or buildings, parking lots, and green spaces forming a university campus (Pruvost and Mooney 2017).

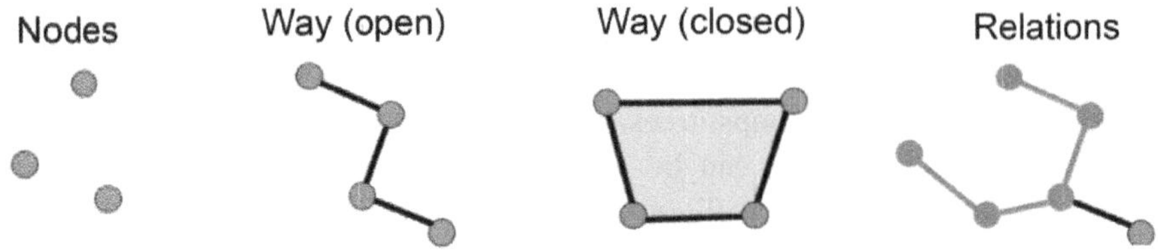

Figure 3.2. *Elements of OpenStreetMap's geographic data model. For a color version of this figure, see www.iste.co.uk/joliveau/geovisualization.zip*

From a semantic perspective, OSM data also differ from traditional vector geographic data, which are characterized by an attribute table composed of fields. Within OSM, nodes, ways and relations are semantically qualified by one or more tags that describe each geographic object. These attribute tags are based on the association of a key and a value. The key specifies the type of object being mapped, and the value provides more detailed information about its properties. For example, a highway, modeled as a series of paths (a set of connected nodes), will be semantically qualified by the key "highway", which identifies the group of roads, and the value "motorway", which specifies the type of road. This geographic object can also be enriched with other tags to provide more information about its characteristics (such as its name, maximum speed limit, number of lanes, surface, operator, etc.). Figure 3.3 illustrates this mechanism of multiple and complementary attributes on three classic objects from the OSM database.

Figure 3.3. *Examples of tags for three types of OSM map objects*

Contrary to geographic data produced by public or private actors, OSM semantics rejects the concept of a fixed, top-down ontology directed by experts (Ballatore and Money 2015). The design of the attribute model is not fixed and evolves collaboratively, according to needs and uses. This mechanism for tagging mapped objects is free, imposing no rules or limitations on the number or content of attributes used, which makes it flexible and scalable (Mocnik et al. 2017). All the tags in OSM constitute a folksonomy[1] co-constructed by contributors (Mocnik et al. 2018). This semi-structured attribute classification allows contributors to create any new attribute to describe objects they consider interesting to map. As of May 2025, there are more than 3.7 billion tags using 181 million different tags, with an average of three tags per node, two tags per way and four tags per relation (Figure 3.4).

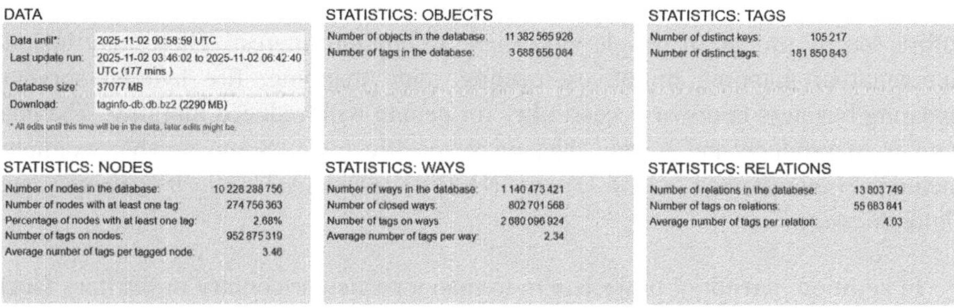

Figure 3.4. *Examples of tags for three types of OSM map objects (source: OSM)*

However, this wide variety of tags does not mean that the production of attribute information is anarchic (Hayat 2019). To make the data usable within GIS software and accessible to a diverse audience, the community works collectively to establish a common vocabulary for standardizing this folksonomy to a minimum. Although the ethic of contributor freedom is strong within the project, contributors follow a set of "informal" standards (Plennert et al. 2019). There is a list of references available on the OSM wiki that lists commonly used keys and values and provides a complete description of them. This process of standardizing the semantics of mapped objects on a global scale represents a major challenge in the development of the project for the coming years (Hall et al. 2018).

---

1 A folksonomy, or personal indexing, is a decentralized, spontaneous collaborative classification system based on indexing carried out by non-specialists.

3.3. A decentralized, organized and diverse community

3.3.1. *OSM: an ecosystem of projects and initiatives*

OSM is an ecosystem that brings together citizens, associations, governments, NGOs and private companies with an interest in mapping, but from very different perspectives. Some participants focus on crisis mapping, others on creating maps in regional languages, and still others on developing technical tools (Ramm et al. 2011; Glasze and Perkins 2015). Thousands of projects revolve around the OSM ecosystem, which is not a single project but rather a "community of communities" (Anderson et al. 2019).

On the one hand, there are a number of mapping projects that are collaborative efforts led by groups of people who share a common interest. These may include humanitarian mapping initiatives, cycling route mapping, fire hydrant surveys, updating business hours or accessibility for people with reduced mobility. The goal here is to produce and update data on a specific topic. Some notable examples include *Humanitarian OSM Team* (HOT)[2], *OpenSeaMap*[3], *Wheelmap*[4] and *PowerNetwork*[5].

In addition, territorial projects aim to map a particular country or territory (such as a region or city), but without any real thematic focus. The idea here is to improve and enrich the cartographic coverage of areas of little detail with a set of reference data (roads, buildings, land use, etc.). There are also projects focused on linguistic and cultural dimensions, such as initiatives to translate place names, with the aim of providing maps in regional languages (Catalan, Breton, Occitan, etc.).

The OSM project also gives rise to multiple technical initiatives focusing on both the database (cleaning, editing, distribution, quality control, etc.) and related services (map rendering, route calculation, geocoding, mobile applications, etc.). Beyond data production, OSM is also an open-source development project where anyone can create and improve tools and features (Plenert 2018). The developers are

---

2 HOT coordinates the production and distribution of open-source mapping resources to support humanitarian assistance efforts around the world (Herfort et al. 2021).

3 *OpenSeaMap* aims to produce nautical information useful to navigators for coastal tourism (beacons, buoys, port information, etc.).

4 *Wheelmap* is an initiative aimed at assessing and mapping the accessibility of places for people with reduced mobility (Mobasheri et al. 2017).

5 *PowerNetwork* aims to map energy facilities around the world (power lines, pylons, substations, production sites, pipelines, etc.).

characterized by their extensive professional experience and advanced skills in database management, software development and system administration. This point highlights that, although the project relies on a community of amateurs, a substantial portion consists of experts and professionals in information technology, cartography and geomatics (Yang et al. 2016).

To conclude this subsection on initiatives related to OSM, it is important to emphasize that one of the specific features of the project compared to other online communities is its strong human and social dimension. In addition to online activities, the community places great importance on interaction and conviviality by regularly organizing meetings between contributors around the world (Hristova 2013). These events can take several forms, such as "cartoparties" for collecting data in the field; "mapathons" for collectively mapping certain areas using imagery; or various more or less formal meetings organized by the OSM Foundation and national and local branches to exchange ideas and organize.

3.3.2. *A contributor-based community*

OSM's main strength lies in its community of contributors and developers, which is large, organized and quite diverse in terms of motivations and skills. Like any online contribution initiative, OSM is characterized by highly uneven participation among contributors (Haklay 2016). This particular dynamic reflects the 90-9-1 rule, according to which less than 1% of participants in an online project contribute actively, 9% participate occasionally and 90% are passive and hardly ever contribute (Begin et al. 2018). Regarding contributor profiles, empirical studies on the subject reveal that the vast majority of participants are male, educated (higher education graduates) and work in the fields of IT, cartography or urban planning (Budhathoki and Haythornthwaite 2013; Duféal and Noucher 2017). Analysis of the age distribution of contributors shows that most contributors are between 20 and 40 years old, and about 20% of cartographers are 40 or older (Coleman et al. 2009).

The motivations that drive contributors are manifold, reflecting the rich variety of forms of engagement, diverse profiles and multiple objectives. Several studies conducted in recent years have provided a better understanding of the factors that motivate contributors to get involved in the OSM project (Budhathoki et al. 2010; Budhathoki and Haythornthwaite 2012; Loftian et al. 2020). On the one hand, these studies reveal a set of intrinsic motivations centered around the contributors, where action is driven solely by the interest and pleasure that the individual finds in it, without any expectation of reward. This may be altruism, activism, leisure, fun or personal fulfillment. There are also extrinsic motivational factors, focused on the contributor and/or the project, where action is determined by motives external to the individual, such as reputation, career, peer recognition, integration into a community or financial reward.

3.3.3. *A community organized around a variety of tools*

Unlike GIS software or web-mapping platforms, there is no single interface for using or contributing to OSM. Instead, a multitude of tools, platforms and applications – including editors, wikis, forums and APIs – are available. The complexity of OSM's technical infrastructure makes it a "black box" that is difficult to understand (Glasze and Perkins 2015; Plennert 2018). Architecturally, OSM is based on a set of technical components related to management (databases, servers, wikis), editing (integrated editors, dedicated software), map rendering (2D and 3D visualization services, tile streams, stylistic rendering interfaces) and data distribution (download platforms, APIs).

In practice, people who interact with the OSM infrastructure do so mainly via editing interfaces that allow them to modify or extend the map coverage (Figure 3.5). Several data editing tools coexist. They offer varying degrees of functionality and are designed for all types of users, such as iD (basic online editor), which allows novices to make simple and quick contributions, or JOSM software, which offers experts a multitude of options for checking, modifying and completing the database (Muttaqien et al. 2018).

In addition to tools for creating and editing maps, the OSM community has a wide range of solutions dedicated to discussion, communication and contributor organization (forums, instant messaging, mailing lists, wikis, etc.). These tools are a very important pillar of collaborative online work outside of mapping, as they enable collective work, exchange and discussion (Neis and Zielstra 2014; Palen et al. 2015). The OSM Wiki, a central component of the OSM ecosystem, is a website that centralizes technical documentation, user guidelines, help resources and promotional content related to the project. Based on MediaWiki content management software, it works in a similar way to Wikipedia, with all content produced and modified collaboratively by community members (Mocnik et al. 2018). The OSM wiki centralizes more than 250,000 pages in some 40 languages. These pages list, document and detail the entire folksonomy, all related projects, contributor groups and organized events, as well as the technical building blocks of the project.

Figure 3.5. *OSM iD editor interface (source: OSM). For a color version of this figure, see www.iste.co.uk/joliveau/geovisualization.zip*

3.3.4. *A decentralized and multi-level project governance*

There is no straightforward answer to the question of who manages and controls OSM. The project has neither a manager nor paid staff. The genetic code of OSM lies in its community, which results in a horizontal and decentralized form of project governance guided by a set of shared norms and practices (Glasze and Perkins 2015). To provide legal status to what might otherwise appear as an informal initiative, the OSMF (based in the United Kingdom) serves as the project's legal entity, ensuring its continuity and growth. This international non-profit organization supports but does not control the project. Unlike Wikipedia, which has sought to manage community growth through an increasingly complex set of rules and a hierarchical system of participation, OSM's governance structure has remained largely unchanged since its inception (Perkins 2014).

The missions and responsibilities of the OSMF remain relatively limited. The Foundation serves as the custodian of the project's computer servers, raises funds, organizes the annual global conference and coordinates thematic, technical and local working groups. The OSMF has approximately 2,000 paying members – including individual contributors, associations, public institutions and companies – who elect a board of directors each year. This board of seven members is responsible for determining the project's overall strategic direction. In addition, the Foundation has an Advisory Board, a consultative body that the board of directors may turn to when making significant decisions. The Advisory Board may also bring forward its own issues or suggestions to the Foundation's leadership. It is composed of one representative from each of the most prominent corporate members, and one representative from each local chapter formally recognized by the OSMF.

To expand its activities internationally, the OSMF has, for several years, been developing and structuring a network of national and regional branches through the establishment of local chapters. Distributed across the globe, these non-profit legal entities operate at the country or regional level and serve as local representatives of the OSMF in their interactions with governments, businesses and the media. As of early 2024, the Foundation had recognized 18 official local chapters, along with 14 unaffiliated groups actively contributing to the development and promotion of the project across five continents (Figure 3.6).

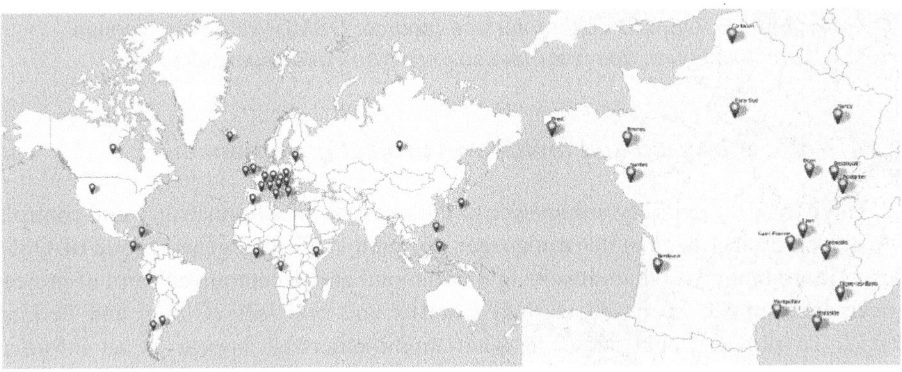

Figure 3.6. *Location of local chapters around the world and contributor groups in France (source: OSM)*

In France, for example, OpenStreetMap France (OSM-FR) is a non-profit association created in 2011 with the aim of promoting the OpenStreetMap project. OSM-FR's missions focus on technical aspects, communication and relations with local authorities (Lambert and Zanin 2012). At the local level, there are also groups of contributors who coordinate at the regional or city level to organize various events (map parties, meetings, project promotion, training, etc.). However, some local groups of contributors operate independently without any real link to OSM-FR, such as OSM-Guyane, whose active members do not fully identify with the issues discussed by the national entity (Noucher 2023).

The organizational ethos of OSM is often characterized as open, democratic and anti-system, insofar as all contributors are regarded as equals. The initial modes of organization among participants resemble a "do-ocracy", in which each contributor acts autonomously and retains full decision-making authority over their own activities (Perkins 2011). In practice, however, the project functions as a hybrid system that combines elements of do-ocracy, meritocracy, technocracy and bureaucracy (Glasze and Perkins 2015). Governance is primarily do-ocratic, insofar as decision-making power resides with those who take action. It is also meritocratic, as volunteer contributions are rewarded with collective recognition. At the same time, it exhibits bureaucratic features, including a certain hierarchy among contributors, a division of responsibilities, standardized procedures and formalized project documentation. Finally, OSM's governance can be described as technocratic, since technical and software development skills are the most valued and recognized forms of expertise within the community (Haklay 2013; Plennert et al. 2018).

The social organization of OSM is therefore not without limitations, as it does not fully guarantee balance and diversity in its governance and can raise issues of representativeness or conflicts of interest. A small group of "elites" consequently makes the most important decisions and shapes the direction and future course of the project. Reflecting this situation, the Foundation's board of directors has gradually become unbalanced. In recent years, there has been an overrepresentation of North American members and a strong emphasis on humanitarian issues (Yang et al. 2024).

3.4. OSM and the professional sphere

Beyond being a project driven by enthusiasts, OSM has become increasingly professional over the years and occupies a growing role in the practices of public, private, academic and non-profit actors. Interactions between the collaborative mapping project and professionals are becoming more frequent and diverse, including data integration and updating, partnerships, technical developments and financial contributions. In this section, we present and discuss the different forms of interaction, their modalities and the underlying issues, depending on the type of actors involved (public sector and private companies).

3.4.1. *Forms of partnership between OSM and the public sector*

A growing number of local authorities, government agencies and parapublic organizations are using OSM not only as a source of information but also as a platform for collectively producing geographic data (Haklay et al. 2014; Jeannot and Cottin-Marx 2022). These new forms of production and updating of institutional geographic data by non-specialists are now taking place in a variety of ways and to varying degrees (Figure 3.7).

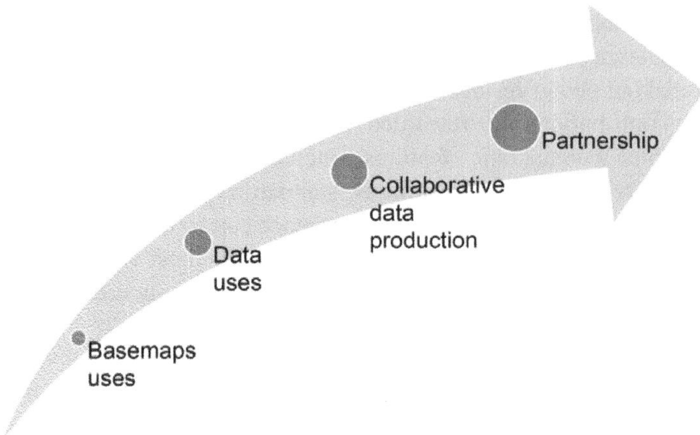

Figure 3.7. *The different levels of interaction between OSM and the public sector*

For many public sector organizations, the main use of OSM is to access the base maps produced by the community for various purposes, such as integrating OSM renderings into their applications. The aim here is to find an alternative to paid base map services such as Google Maps. In practice, public organizations can either use base map feeds hosted by the Foundation or host OSM data for a given territory internally and redistribute it free of charge to their partners. Some national administrations (such as Etalab in France) or regional ones (Brittany, Grand Est, Catalonia) offer map tile feeds to display OSM data in web applications free of charge and without limits.

At a more advanced level, public actors can use OSM data to enrich and supplement their own business databases. Internally, OSM is mainly used to feed GIS systems or to verify and supplement government databases (Fairhurst 2020). The main focus of this practice is the road network; many local authorities and ministries now integrate OSM data into their database updating processes (the city of Antwerp, the Norwegian Ministry of Transport, the Helsinki Public Transport Authority, Portland transport agencies, etc.).

Public actors can also mobilize the technical infrastructure of OSM and its community to produce and update data on specific topics (tourism, transportation, accessibility). Institutional commitments to OSM often take the form of organized contribution activities (workshops, map parties) aimed at enriching the database for a specific region. Many initiatives focus on tourism, involving the mapping and detailed documentation of accommodation, restaurants, leisure activities, gas stations and vending machines. There are also projects focusing on accessibility, where local authorities organize collaborative mapping of routes suitable for people with reduced mobility. In France, for example, the public transport operator Île-de-France Mobilité has mobilized the OSM infrastructure to digitize and characterize cycling facilities. As for the emergency services, many organizations such as the SDIS (fire and rescue services) use OSM through the OSMHydrants project to identify, classify and update their databases on fire hydrants, water tanks and pumping stations.

The community can also establish closer collaboration by using OSM data and at the same time participating in its ongoing and organized updating. In France, as part of its open data initiative, Montpellier Méditerranée Métropole has, for example, established cooperation with the local OSM community to update data relating to its 31 municipalities (Plantin and Valentin 2013). On the one hand, local government officials contribute to OSM on an ongoing basis on various topics, and, on the other hand, OSM data are integrated into the local government's open data portal to offer re-users a wide range of territorial data.

The most advanced form of interaction between OSM and public actors takes the form of partnerships. In France, for example, the French national railway company (SNCF) has established a partnership with OSM-FR to map the interiors of major passenger stations, focusing on facilities, services and accessibility (Denis 2019). More recently, in 2021, the national electricity distribution network operator ENEDIS and OSM-FR signed an agreement representing a further step toward the exchange and co-production of large-scale spatial data. The availability of high-definition imagery enables contributors to provide more accurate information on the locations and characteristics of electrical infrastructure.

One of the drivers of these new forms of institutional geographic data co-production lies in the development of trusting relationships between public bodies and local associations and businesses working with OSM. The awareness-raising and communication work carried out by national chapters and local contributor groups is key to developing and sustaining collaborations with the public sector. However, the integration of OSM and, by extension, data produced by non-specialists into public sector databases raises many questions in terms of quality, licensing and sovereignty (Saunders et al. 2012; Johnson 2017). This new paradigm in public data production takes time and still raises concerns among public actors (Khan and Johnson 2020). On the one hand, they have clearly identified the opportunities offered by OSM, but, on the other hand, they remain fairly cautious in their collaboration with the project. National mapping agencies are conspicuously absent from these new forms of partnership, preferring to develop their own platforms for reporting and updating their geographic databases (Olteanu-Raimond et al. 2017).

3.4.2. *Forms of partnership between OSM and the industry sector*

The landscape of large companies, and more specifically digital platforms that produce and use geographic data, has changed radically over the past 10 years (Leszczynski 2012). At the heart of these developments are the major tech companies, for whom maps have become an essential part of their services, occupying a central role in the evolution of digital cartography (Dalton 2015). Amid the restructuring among players – acquisitions, mergers, and partnerships – and the emergence of new approaches to spatial data production, such as the spread of crowdsourcing, digital giants have clearly recognized the advantages of leveraging and updating OSM data to enhance and improve their services (Sarkar and Anderson 2022).

Unlike public actors, who primarily use OSM data without necessarily contributing to it, large corporations are playing an increasingly important role in data production and technical developments within the project (Yang et al. 2024). Facebook, Microsoft, Apple, Amazon and Uber have actively collaborated on the advancement of OSM, setting aside their specific objectives to provide their users with increasingly accurate and up-to-date maps (Michel and Schröder-Bergen 2022). The number of corporate contributors in the OSM project grew from around a dozen to more than 1,000 in 2019 and 2,300 in 2020 (Anderson and Sarkar 2020). From funding the foundation to large-scale data production and the implementation of new tools, the forms of involvement are becoming increasingly numerous and diverse (Anderson et al. 2019). The objects mapped – roads, buildings, businesses, etc. – and the geographic coverage also vary from one company to another. However, there is a stronger focus on road networks in regions where these companies are most active.

Microsoft was the first major digital company to collaborate with OSM through its Bing Maps service. In 2010, the multinational made its aerial imagery service available free of charge within its editing tools to facilitate contributions. That same year, it hired OSM founder Steve Coast as lead architect for its mobile mapping service. In 2017, following the digitization of buildings in the United States, Canada and Australia, more than 125 million buildings were added to the OSM database. More recently, the company has also made street-level images from its StreetSide service available to the community for editing objects such as street furniture, signage and building materials.

Apple chose OSM in 2012 to power the base map for its iPhoto photo management service. Within 2 years, the company's contributions grew from fewer than 2 million edits to nearly 15 million. An organized editing team composed of Apple employees contributes daily to enriching the database across approximately 40 countries. The company is also involved in new developments related to data verification and quality control, such as the Atlas tool, which analyzes the topology of OSM objects and generates task lists for contributors.

Meta (Facebook) has been using OSM data since 2018 to support several of its services, including mapping and location sharing. The multinational company actively participates in enriching OSM data, particularly in geographic areas with limited coverage. For example, Facebook launched the mapwith.ai platform, which uses artificial intelligence to automatically detect roads and paths from high-resolution satellite imagery (Schröder-Bergen et al. 2021). Once these objects have been identified by the algorithm, contributors simply validate or correct the predictions, saving them the tedious task of manually adding objects. One of the project's goals is to rapidly map areas affected by events requiring humanitarian intervention, where data is lacking. The first territory covered by mapwith.ai was Thailand, where nearly 500,000 km of roads were added to OSM in less than 18 months. Since then, Indonesia, Haiti and Malaysia have also been covered using this approach.

Unlike digital giants, for whom geographic data is just one component of their commercial offerings, other large companies specializing in web mapping, such as Mapbox, Stamen or MapTiler, have built their business models around leveraging OSM data. These companies, which monetize various mapping services based on community-produced data (basemaps, geocoding and route planning) also actively participate in updating the data and contributing to technical advancements within the OSM project.

The emergence of new actors providing detailed and customizable basemaps for commercial use has introduced new dynamics in the way large private companies engage with OSM data (Nass et al. 2019). Numerous online platforms and services have migrated from Google Maps to these new providers, and by extension to OSM, without necessarily contributing directly to the project (Figure 3.8). Notable examples include the social media platforms Snapchat (through its Snap Map) and Foursquare, the game developer Niantic (Pokémon Go) and the car manufacturer Tesla.

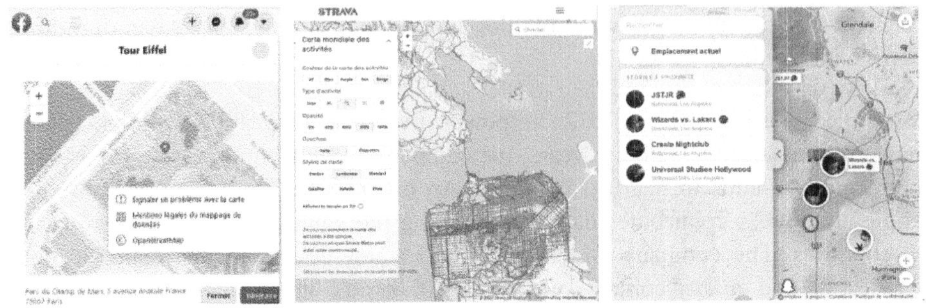

Figure 3.8. *Online map interfaces for Facebook, STRAVA, and Snapchat platforms based on data from OSM. For a color version of this figure, see www.iste.co.uk/joliveau/geovisualization.zip*

Traditional companies specializing in map publishing or online services, such as Michelin and Mappy, have adopted different strategies. For these "historic" private mapping operators, OSM data are a source of complementary data, but their interactions with the project remain fairly limited (Hayat 2019).

OSM data play an increasingly central role in the daily lives of millions of people, often serving as an alternative to the basemaps traditionally provided by Google Maps. The Palo Alto–based company remains conspicuously absent from these new partnerships between private firms and OSM. Instead, Google has chosen to retain full control over the production and updating of its spatial data by developing its own contribution tools and coordinating a collaborative mapping communities, such as Map Maker (discontinued in 2013) and the Waze Map Editor, which operates according to principles similar to those of OSM (Hind and Gekker 2014).

3.5. Some major challenges for OSM in mapping and geomatics

To conclude this overview of OSM, the final section adopts a forward-looking perspective by examining a series of technical and political issues shaping the project's future. After two decades of existence, OSM faces numerous and diverse challenges that span data management, technological tools, partnerships and community dynamics. This section highlights three areas of particular significance for the project's development in the coming years: data quality and vandalism, 3D modeling and indoor mapping and the notion of the digital commons.

3.5.1. *The quality of OSM data and map vandalism*

The quality of OSM data has received considerable attention and scrutiny from the scientific community as well as from public and private stakeholders (Senaratne et al. 2017; Moradi et al. 2022). The prominence of this issue reflects the challenge of ensuring the usability of data produced by non-specialists in a variety of operational contexts, including public policy and commercial services. With contributions from a diverse range of users, ensuring quality in all dimensions is critical, since non-expert contributors may introduce geometric (particularly topological) or semantic errors, including inappropriate tags on mapped objects.

Firstly, data produced by the OSM community appears to be of high quality, both spatially and in terms of attributes. The collaborative editing process for geographic features yields highly accurate, topologically consistent and semantically rich data. Thanks to ongoing supervision of contributions and frequent updates, OSM is a data source comparable to, or in some cases surpassing, the quality of institutional and commercial databases, and often more up-to-date (Girres and Touya 2010; Zielstra and Zipf 2010). However, the primary limitation lies in the comprehensiveness of coverage, which varies substantially across regions. Data quality tends to be higher in densely populated areas and lower in sparsely populated or rural regions (Zhou, 2018; 2022). This contrasts with geographic databases produced by national mapping agencies, which maintain consistent levels of completeness across all territories. Additionally, heterogeneity and semantic complexity can pose challenges to data quality due to the lack of harmonization (Barron et al. 2014).

In addition to uncertainties inherent in the internal quality of its data, OSM is also vulnerable to deliberate manipulation or degradation of geographic information. This phenomenon, commonly referred to as cartographic vandalism, can arise from a variety of motivations and take multiple forms, including playful, ideological, fantastical or even artistic acts (Ballatore 2014). Within the OSM context, vandalism is defined as the intentional violation of community-established standards (Neis et al. 2012). Such acts may involve copyright infringement, the addition of graffiti, the deletion or damage of objects or the use of automated scripts (bots). More specifically, vandalism can manifest as the application of inappropriate attributes, the deletion of existing objects, the creation of fictitious features or the modification or removal of objects in contested or disputed areas, as illustrated in Figure 3.9 (Singla et al. 2014).

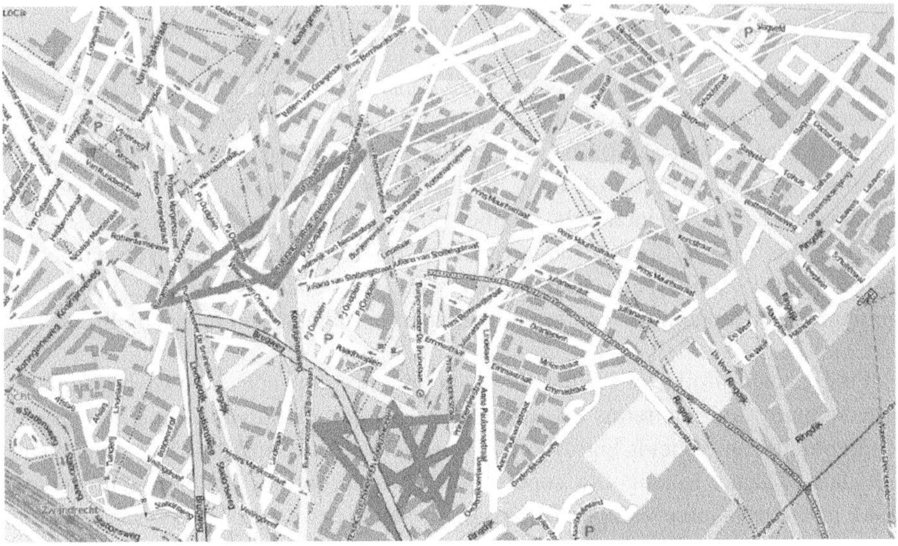

Figure 3.9. *Example of cartographic vandalism within the OSM database (source: Neiss et al. 2012). For a color version of this figure, see www.iste.co.uk/joliveau/geovisualization.zip*

Vandalism represents a persistent challenge within OSM, with numerous documented incidents demonstrating its impact on data integrity. For example, in 2012, Google employees were implicated in the deletion of OSM data and the reversal of one-way street designations. Similarly, in 2018, the OSM editor was exploited to introduce anti-Semitic content onto the Snapchat social network map for several minutes (OSM Wiki 2018). Beyond these isolated cases, politically motivated vandalism frequently targets contested territories, including regions along the Chinese border, Ukraine and Palestine (Bittner 2017). The integration of OSM data into the mobile game Pokémon Go has also produced instances of vandalism, as players manipulated the collaborative map to create false data and gain an unfair advantage (Juhász et al. 2020).

The OSM community's capacity to address data quality and vandalism can be analyzed from two complementary perspectives. First, OSM is not simply a cumulative database in which contributors independently add new elements, it functions as a cartographic wiki, with a shared objective of collaboratively producing rich, high-quality geographic content. As a communicative and collaborative system, OSM fosters community self-regulation and individual self-assessment through continuous interaction among members, promoting the construction of shared meaning (Tourné 2008). In practical terms, this means that if an error occurs – whether intentional or accidental – there is a strong likelihood that another community member will promptly intervene to correct it constructively. Consequently, acts of vandalism are typically removed within minutes or hours and are generally disruptive only on a sporadic basis (Juhász et al. 2020).

Beyond self-regulation by contributors, the OSM community has established protocols and technical solutions specifically designed to ensure data quality and manage vandalism (Singla et al. 2014). These include features and tools – such as editor notes – that allow contributors to flag potential errors. Additionally, automated error detection tools are available to identify inconsistencies and gaps in the data. While manual review by contributors remains the community's primary approach, there is a growing emphasis on developing AI-driven methods to automatically detect acts of vandalism and suspicious user profiles (Neis et al. 2012; Truong et al. 2020).

3.5.2. *Toward convergence between 3D modeling and indoor mapping*

The development of 3D modeling, which has experienced a technical and practical revolution in the GIS field over the past few years, poses another significant challenge for OSM (Knoth et al. 2017). In an era marked by the widespread adoption of digital models, digital twins, and Building Information Modeling (BIM), the transition from 2D to 3D mapping is a critical focus for the community actively engaged in this area. While OSM was originally designed for two-dimensional mapping, its data model has since evolved. Beginning in 2012, discussions among specialists and enthusiasts led to the creation of descriptive frameworks for capturing the 3D attributes of buildings within the database (Goetz and Zipf 2013).

The Simple 3D Buildings tagging system allows, for example, in addition to traditional information about buildings (address, use, etc.), other tags to be added to characterize their volume and shape in greater detail (height, number of floors, roof shape, color, materials, etc.). In addition, the more complex 3D tagging system offers to map elements on a larger scale: buildings, entrances, balconies, windows or exterior staircases. This form of micro-mapping also applies to objects other than buildings, allowing the height or orientation of bridges, statues, streetlights or trees to be specified.

Figure 3.10. *3D rendering of the Rennes 2 University campus based on OSM data (source: F4Map). For a color version of this figure, see www.iste.co.uk/joliveau/geovisualization.zip*

All this new geographic data on building volume and architectural details can be used to generate realistic 3D geovisualizations, as illustrated in Figure 3.10. Depending on the coverage, it is thus possible to create large-scale digital urban models at low cost (Over et al. 2010; Goetz 2013). The OSM technical spectrum includes several 3D engines for displaying buildings on a web map. Some of these are open source, such as OSMBuildings, while others are proprietary, such as F4Map.

In addition to 3D building modeling, indoor mapping is another important challenge for OSM in the coming years (Goetz and Zipf 2011; Chen and Clarke 2020). The aim of indoor mapping is to provide accurate spatial data that can be used to find one's way around all levels of a building (rooms, corridors, access points, elevators, stairways, etc.). This type of approach is particularly useful in large complexes (shopping malls, airports, train stations, universities) for finding specific locations, navigation or accessibility for people with reduced mobility.

Indoor mapping poses a major technical challenge for OSM, namely, the management of geometry levels within buildings. Since 2014, the OSM community has had a set of stabilized and documented tags to model and describe these complex configurations (floors, rooms, corridors and staircases) in detail. Simple Indoor

Tagging, as a tagging scheme for indoor mapping, allows the modeling of building floors, their internal layout and the connections between different levels (stairs, elevators and escalators). There are numerous indoor mapping initiatives around the world that have used the OSM infrastructure to digitize the internal layout of public buildings (train stations, shopping malls, museums, university campuses, etc.). To make the most of this particular type of geographic data, the community has developed indoor mapping rendering and navigation tools such as OpenLevelUp and OpenIndoor (Figure 3.11).

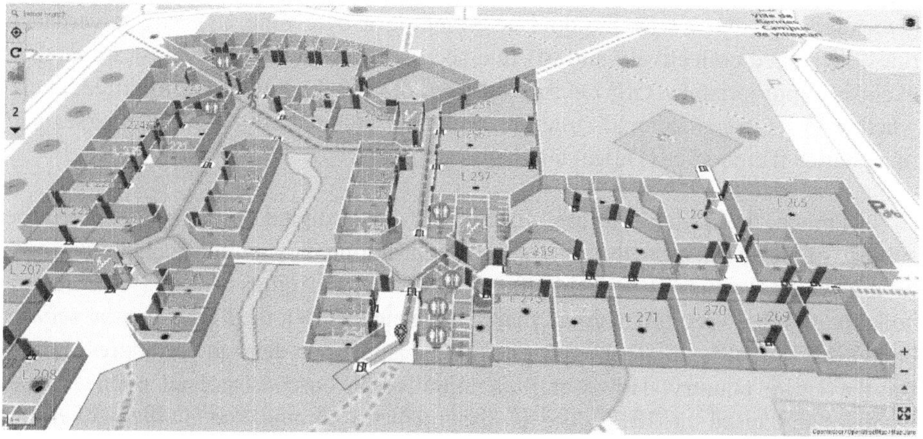

Figure 3.11. *Indoor mapping based on OSM data for a building at Rennes 2 University (source: OpenIndoor). For a color version of this figure, see www.iste.co.uk/joliveau/geovisualization.zip*

The issues of 3D modeling and indoor mapping are still approached in different ways within OSM. To respond to the challenges posed by the convergence of these two trends and to align with new modeling approaches, particularly BIM, the community is working on a new, more integrative model called Full 3D Building (F3DB). The characteristics of F3DB are based on a set of features that describe 2D, 2.5D and 3D data with separate renderings of the internal and external structures of buildings, level by level. The aim is to link OSM to new fields of geomatics applications such as BIM or CIM (Tang et al. 2019) and to make OSM data interoperable with civil engineering and building standards, such as IFC or CityGML formats (Goetz 2013). This type of approach also enables the development of new mapping services, such as 3D visualization of building interiors and indoor route calculation.

3.5.3. *The OSM project as a digital (geo)commune?*

To conclude this chapter, it is important to return to the concept of the digital commons, which is gaining increasing attention in both academic and policy debates. A digital commons can be understood as a digital resource that is created, managed and maintained collectively by a diverse community of actors, and governed according to jointly developed rules designed to preserve its open and shared nature (Dulong de Rosnay and Stalder 2020). Shulz (2021) further defines a digital commons as (1) a community of producers and users who manage, (2) a digital resource with the goal of enriching it over time, (3) with governance rules developed collectively and (4) free access that is protected against attempts at exclusive appropriation. OSM embody a form of digital commons at multiple levels, whether in its organizational structure, technical architecture or the discourses and imaginaries it embodies (Duféal and Noucher 2017).

In France, the question of establishing OSM as a digital commons has been the subject of debate within the community and among key public stakeholders for several years. Launched in 2014, the Open National Address Database project, a collaboration between the national mapping agency (IGN), the French postal service and OSM, sought to create the most comprehensive open database of address points for the entire country. However, this initiative to co-construct an open address database collapsed in 2019 due to disagreements among stakeholders over the choice of data license (Jeannot and Cottin-Marx 2022). According to Shulz (2021), the abandonment of the digital commons model for the national address database illustrates how the state exercises a particular form of informational power, manifested in the technical restriction of citizen contributions while legally facilitating data reuse by commercial actors.

Following this setback, the national mapping agency recently launched a new policy with a stronger focus on open data and free software, developing a strategy around the concept of "Geo-commons". For the IGN, which is seeking to redefine its role in a radically transformed ecosystem, OSM is frequently cited and actively promoted. On the public side, this has generated significant communication and publicity, while community expectations remain high. The local chapter, OSM-FR, has clearly positioned itself within this Geo-commons perspective by attempting to coordinate the various stakeholders through communications, meetings and a dedicated forum to agree on strategies to implement and develop new initiatives. Since 2021, the Panoramax project, which aims to build a free and open database of immersive views as an alternative to Google Street View, has been integrated into a state public policy initiative, called "Geo-commons Fabric". Managed and funded by the IGN, the project, which encompasses over 70 million photographs, actively involves the OSM community in its technical development and governance. However, each actor maintains its own technical authority and distributes data under different licenses (Open Government Licence and Open Database License). This example, like the aborted BANO project, clearly illustrates that cooperation between OSM and a state actor is complex and raises sociotechnical issues that often diverge – or even conflict – from political, economic and ideological perspectives.

The growing involvement of major technology companies in the development and governance of OSM also raises questions about the project's future as a digital commons. Of the 77 organized editing teams recognized by the community, more than two-thirds are now supported by large corporations such as Uber, Amazon, Microsoft and Apple. These companies play an increasingly significant role in technical development, the editing of specific data layers (particularly roads) and the financing of the Foundation. As of March 2019, nearly 17% of the global road network had been edited and updated by teams supported by major digital companies, with this proportion exceeding 90% in certain countries, such as Thailand and Vietnam (Anderson et al. 2019; Sarkar and Anderson 2022). This rapid expansion in the number of corporate editors on the platform, and consequently in the volume of data produced, has become a growing source of controversy within the OSM community.

These tech giants promote and showcase the work of contributors while acting as facilitators, enabling organizations to integrate OSM building blocks into large-scale projects (Quest and Bonnet 2016). At the same time, however, these new entrants are playing an increasingly influential role on the OSM Foundation's board and are transforming the processes and tools used by contributors. In principle, the increase in the volume of mapped objects appears beneficial to the project's development. Yet, the production methods employed by these actors – often relying on AI or paid labor – can diverge from, or even contradict, the ideology of a digital commons (Michel and Schröder-Bergen 2022). For example, Meta imported a substantial amount of erroneous data into Egypt, using the mapwith.ai tool, disregarding the code of conduct for automated edits. The data quality was so poor that it was ultimately removed from the database. Furthermore, the increasing reliance on volunteers by large companies to produce geographic data at lower cost raises concerns about a form of "data extractivism" (Schröder-Bergen et al. 2021). This situation calls into question the roles of civil servants, employees and volunteers in the collaborative production of geographic information (Jeannot and Cottin-Marx 2022).

It is noteworthy that several of the large companies involved in OSM have recently decided to develop their own geographic data production projects to support their services and applications. Launched in 2022, the Overture Maps Foundation (OMF) project, which brings together Amazon, Microsoft, Meta, TomTom and the Linux Foundation, produces several freely accessible global vector geographic databases covering roads, buildings and points of interest (Ballantyne and Berragan 2023).

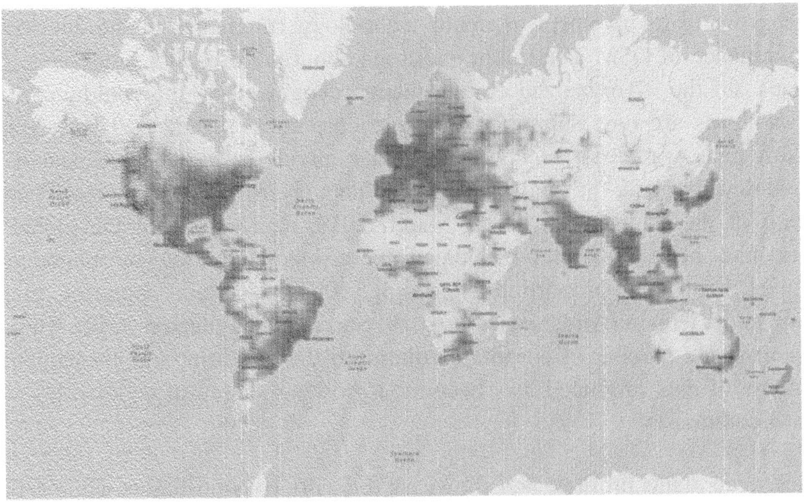

Figure 3.12. *Spatial coverage of Overture Map's 69 million POIs (source: OMF). For a color version of this figure, see www.iste.co.uk/joliveau/geovisualization.zip*

However, this open spatial data project, which incorporates some data from OSM, differs significantly from OSM in terms of production methods (automatic data generation using AI), technical architecture (centralized on Amazon and Microsoft clouds), data availability (particularly licensing) and governance (a consortium of tech companies). Overture Maps is a global mapping initiative that aims to produce reliable, comprehensive and harmonized geographic data, rather than a community of individual editors mapping freely according to a few rules and a shared ideology. This initiative positions itself as an alternative to the dominance of Google Maps in the field of geographic data, rather than as a replacement for OSM.

The absence of the OSMF from the Overture Maps project – which was either unable or unwilling to develop and formalize more extensive partnerships with the private sector – reinforces its relegation to the role of a data provider. In recent years, the OSM community has not actively pursued reforms in governance, licensing or the usability of its data by operational actors. Nevertheless, Overture Maps does not signal the decline or end of OSM. On the contrary, this project, led by digital giants with a voracious appetite for geographic data, challenges the OSM community, which is striving to reinvent itself and evolve to secure its place within a constantly changing geospatial ecosystem.

3.6. Conclusion

As we have seen throughout this chapter, OSM is a rich and complex collaborative mapping project from both technical and organizational perspectives. It is simultaneously a freely editable and usable geographic database, a diverse community of individuals and organizations collaboratively building the map in a decentralized manner, and a suite of technical tools supporting the entire data lifecycle. This sociotechnical ecosystem, which encompasses thousands of projects worldwide, is multifaceted and inherently distributed and decentralized. Its governance is shared, its operation both meritocratic and technocratic, and its codes and culture are centered on openness and freedom.

With regard to OSM, geospatial professionals have gradually shifted from a position of disparagement to one of cautious interest, followed by increasingly significant engagement. The work of contributors, now more organized, efficient and recognized, has positioned the OSM ecosystem as a key actor in the geospatial sector. In recent years, the project has been marked by the professionalization of practices and a degree of institutionalization that question public sector organizations, private companies and academic.

The OSM community faces numerous challenges in achieving maturity, consolidating its technical architecture, securing its foundation, stabilizing governance and establishing its position with respect to public and private stakeholders. Internally, the accuracy and standardization of data – underpinned by quality – remain central and constantly evolving concerns. The development of new domains, such as mobile applications, 3D modeling and indoor mapping, represents significant challenges for the coming years. Externally, the resurgence of the concept of digital sovereignty, drawing in public actors and their broader ecosystems, increasingly shifts the balance in favor of OSM. In the private sector, stakes are high for both small service providers and digital giants with an insatiable appetite for the data produced. Vigilance remains essential to prevent OSM from being appropriated by private entities whose objectives and ethical conflict with the ideological foundations of the project.

OSM celebrated its 20th anniversary, 20 years marked by success and the recognition of a simple idea: mapping the world together for everyone. There is no doubt that the project's growth and development, now well beyond its adolescence, will continue to transform the digital mapping sector in the years to come. Finally, to conclude this chapter, it is important to remember that the project is open to everyone, from specialists to novices, from curious minds to enthusiasts. Every contribution, no matter how modest, helps advance the project. Teachers, researchers and students working directly or indirectly with geographic data all have a role to play in this cartographic adventure.

3.7. References

Anderson, J. and Sarkar, D. (2020). Curious cases of corporations in OpenStreetMap. In *Proceedings of the Academic Track at the State of the Map 2020 Online Conference*. Zenodo.

Anderson, J., Sarkar, D., Palen, L. (2019). Corporate editors in the evolving landscape of OpenStreetMap. *ISPRS International Journal of Geo–Information*, 8(5), 232.

Ballantyne, P. and Berragan, C. (2023). Overture POI data for the United Kingdom: A comprehensive, queryable open data product. *arXiv*. doi: 10.48550/arXiv.2310.18415.

Ballatore, A. (2014). Defacing the map: Cartographic vandalism in the digital commons. *The Cartographic Journal*, 51(3), 214–224.

Ballatore, A. and Mooney, P. (2015). Conceptualising the geographic world: The dimensions of negotiation in crowdsourced cartography. *International Journal of Geographical Information Science*, 29(12), 2310–2327.

Barron, C., Neis, P., Zipf, A. (2014). A comprehensive framework for intrinsic OpenStreetMap quality analysis. *Transactions in GIS*, 18(6), 877–895.

Bégin, D., Devillers, R., Roche, S. (2018). The life cycle of contributors in collaborative online communities – the case of OpenStreetMap. *International Journal of Geographical Information Science*, 32(8), 1611–1630.

Bittner, C. (2017). OpenStreetMap in Israel and Palestine – 'Game changer' or reproducer of contested cartographies? *Political Geography*, 57, 34–48.

Budhathoki, N.R. and Haythornthwaite, C. (2012). Motivation for open collaboration: Crowd and community models and the case of OpenStreetMap. *American Behavioral Scientist*, 57(5), 548–575.

Budhathoki, N.R. and Haythornthwaite, C. (2013). Motivation for open collaboration: Crowd and community models and the case of OpenStreetMap. *American Behavioral Scientist*, 57(5), 548–575.

Budhathoki, N.R. and Nedović–Budić, Z. (2010). An interdisciplinary frame for understanding volunteered geographic information. *Geomatica*, 64(1), 11–26.

Chen, J. and Clarke, K.C. (2020). Indoor cartography. *Cartography and Geographic Information Science*, 47(2), 95–109.

Coast S. (2015). *The Book of OSM*. Createspace Independent Publishing Platform, Scott Valleys.

Coleman, D.J., Georgiadou, Y., Labonte, J. (2009). Volunteered geographic information: The nature and motivation of produsers. *International Journal of Spatial Data Infrastructures Research*, 4, 332–358.

Dalton, C.M. (2015). For fun and profit: The limits and possibilities of Google Maps based geoweb applications. *Environment and Planning A*, 47(5), 1029–1046.

Denis, J. and Marquet, C. (2019). De la production à la maintenance : faire exister collectivement des données géographiques ouvertes. *Questions de communication*, 36, 63–83.

Duféal, M. and Noucher, M. (2017). Des TIC au TOC. Contribuer à OpenStreetMap : entre commun numérique et utopie cartographique. *Netcom. Réseaux, communication et territoires*, 31(1/2), 77–98.

Dulong de Rosnay, M. and Stalder, F. (2020). Digital commons. *Internet Policy Review*, 9(4), 1–22.

Estima, J. and Painho, M. (2013). Exploratory analysis of OpenStreetMap for land use classification. In *Proceedings of the Second ACM SIGSPATIAL International Workshop on Crowdsourced and Volunteered Geographic Information*. ACM, New York.

Fairhurst, R. (2020). *Providing Data to OpenStreetMap – A New Guide for Data Owners*. Oxfordshire County Council and Buckinghamshire Council.

Feldmeyer, D., Meisch, C., Sauter, H., Birkmann, J. (2020). Using OpenStreetMap data and machine learning to generate socio–economic indicators. *ISPRS International Journal of Geo-Information*, 9(9), 498.

Girres, J.F. and Touya, G. (2010). Quality assessment of the French OpenStreetMap dataset. *Transactions in GIS*, 14(4), 435–459.

Glasze, G. and Perkins, C. (2015). Social and political dimensions of the OpenStreetMap project: Towards a critical geographical research agenda. In *OpenStreetMap in GIScience*, Arsanjani, J, Zipf, A., Mooney, P., Helbich, B. (eds). Springer, Cham.

Goetz, M. (2013). Towards generating highly detailed 3D CityGML models from OpenStreetMap. *International Journal of Geographical Information Science*, 27(5), 845–865.

Goetz, M. and Zipf, A. (2011). Extending OpenStreetMap to indoor environments: Bringing volunteered geographic information to the next level. In *Urban and Regional Data Management: Udms Annual 2011*, Rumor, M., Zlatanova, S., LeDoux, H. (eds). CRC Press, Delft.

Goetz, M. and Zipf, A. (2013). The evolution of geo–crowdsourcing: Bringing volunteered geographic information to the third dimension. In *Crowdsourcing Geographic Knowledge*, Sui, D., Elwood, S., Goodchild, M. (eds). Springer, Dordrecht.

Goodchild, M. (2007). Citizens as sensors: The world of volunteered geography. *GeoJournal*, 69(4), 211–221.

Grippa, T., Georganos, S., Zarougui, S., Bognounou, P., Diboulo, E., Forget, Y., Lennert, M., Vanhuysse, S., Mboga, N., Wolff, E. (2018). Mapping urban land use at street block level using OpenStreetMap, remote sensing data, and spatial metrics. *ISPRS International Journal of Geo–Information*, 7(7), 246.

Haklay, M. (2013). Neogeography and the delusion of democratisation. *Environment and Planning A*, 45(1), 55–69.

Haklay, M. (2016). Why is participation inequality important? In *European Handbook of Crowdsourced Geographic Information*, Capineri, C., Haklay, M., Huang, H., Antoniou, V., Kettunen, J., Ostermann, F., Purves, R. (eds). Ubiquity Press, London.

Haklay, M., Antoniou, V., Basiouka, S., Soden, R., Mooney, P. (2014). *Crowdsourced Geographic Information Use in Government.* World Bank Publications, London.

Hall, A., Thebault-Spieker, J., Sen, S., Hecht, B., Terveen, L. (2018). Exploring the relationship between "informal standards" and contributor practice in OpenStreetMap. In *Proceedings of the 14th International Symposium on Open Collaboration.* ACM, Paris.

Hayat, F. (2019). Production des biens communs numériques et usages cartographiques. PhD Thesis, Université de Paris, Paris.

Heipke, C. (2010). Crowdsourcing geospatial data. *ISPRS Journal of Photogrammetry and Remote Sensing*, 65(6), 550–557.

Herfort, B., Lautenbach, S., Porto de Albuquerque, J., Anderson, J., Zipf, A. (2021). The evolution of humanitarian mapping within the OpenStreetMap community. *Scientific Reports*, 11(1), 1–15.

Hind, S. and Gekker, A. (2014). "Outsmarting traffic, together": Driving as social navigation. *Exchanges: The Interdisciplinary Research Journal*, 1(2), 165–180.

Hristova, D., Quattrone, G., Mashhadi, A., Capra, L. (2013). The life of the party: Impact of social mapping in OpenStreetMap. In *Proceedings of the International AAAI Conference on Web and Social Media.* AAAI Press, Cambridge.

Jeannot, G. and Cottin–Marx, S. (2022). *La privatisation numérique : déstabilisation et réinvention du service public.* Raisons d'Agir, Paris.

Johnson, P.A. (2017). Models of direct editing of government spatial data: Challenges and constraints to the acceptance of contributed data. *Cartography and Geographic Information Science*, 44(2), 128–138.

Juhász, L., Novack, T., Hochmair, H.H., Qiao, S. (2020). Cartographic vandalism in the era of location-based games – the case of OpenStreetMap and Pokémon GO. *ISPRS International Journal of Geo–Information*, 9(4), 197.

Khan, Z.T. and Johnson, P.A. (2020). Citizen and government co-production of data: Analyzing the challenges to government adoption of volunteered geographic information. *The Canadian Geographer/Le Géographe canadien*, 64(3), 374–387.

Knoth, L., Mittlboeck, M., Vockner, B. (2017). 3D building maps for everyone – mapping buildings using VGI. In *Societal Geo-innovation: Selected Papers of the 20th AGILE Conference on Geographic Information Science*, Bregt, A., Sarjakoski, T., Lammeren, V., Rip, F. (eds). Springer, Cham.

Lambert, N. and Zanin, C. (2012). OpenStreetMap : collaborer pour faire des cartes. *Mappemonde*, 107(3).

Leszczynski, A. (2012). Situating the geoweb in political economy. *Progress in Human Geography*, 36(1), 72–89.

Liu, X. and Long, Y. (2016). Automated identification and characterization of parcels with OpenStreetMap and points of interest. *Environment and Planning B: Planning and Design*, 43(2), 341–360.

Lotfian, M., Ingensand, J., Brovelli, M.A. (2020). A framework for classifying participant motivation that considers the typology of citizen science projects. *ISPRS International Journal of Geo–Information*, 9(12), 704.

Luaces, M.R., Fisteus, J.A., Sánchez–Fernández, L., Munoz–Organero, M., Balado, J., Díaz–Vilariño, L., Lorenzo, H. (2020). Accessible routes integrating data from multiple sources. *ISPRS International Journal of Geo–Information*, 10(1), 7.

Michel, B. and Schröder–Bergen, S. (2022). The politics of geodata in urban platform capitalism. In *Platformization of Urban Life*, Strüver, A. and Bauriedl, S. (eds). Columbia University Press, New York.

Mobasheri, A., Deister, J., Dieterich, H. (2017). Wheelmap: The wheelchair accessibility crowdsourcing platform. *Open Geospatial Data, Software and Standards*, 2(1), 1–7.

Mocnik, F.B., Zipf, A., Raifer, M. (2017). The OpenStreetMap folksonomy and its evolution. *Geo–spatial Information Science*, 20(3), 219–230.

Mocnik, F.B., Mobasheri, A., Zipf, A. (2018). Open–source data mining infrastructure for exploring and analysing OpenStreetMap. *Open Geospatial Data, Software and Standards*, 3(1), 1–15.

Moradi, M., Roche, S., Mostafavi, M. (2022). Exploring five indicators for the quality of OpenStreetMap road networks: a case study of Québec, Canada. *Geomatica*, 75(4), 1–31.

Muttaqien, B.I., Ostermann, F.O., Lemmens, R.L. (2018). Modeling aggregated expertise of user contributions to assess the credibility of OpenStreetMap features. *Transactions in GIS*, 22(3), 823–841.

Nass, A., Manaud, N., van Gasselt, S., Hare, T. (2019). Towards a new face for planetary maps: Design and web-based implementation of planetary basemaps. *Advances in Cartography and GIScience of the International Cartographic Association*, 1(15), 1–8.

Neis, P. and Zielstra, D. (2014). Recent developments and future trends in volunteered geographic information research: The case of OpenStreetMap. *Future Internet*, 6(1), 76–106.

Neis, P. and Zipf, A. (2012). Analyzing the contributor activity of a volunteered geographic information project – The case of OpenStreetMap. *ISPRS International Journal of Geo–Information*, 1(2), 146–165.

Neis, P., Goetz, M., Zipf, A. (2012). Towards automatic vandalism detection in OpenStreetMap. *ISPRS International Journal of Geo–Information*, 1(3), 315–332.

Noucher, M. (2023). *Blancs des cartes et boîtes noires algorithmiques*. CNRS Éditions, Paris.

Olteanu-Raimond, A.M., Hart, G., Foody, G.M., Touya, G., Kellenberger, T., Demetriou, D. (2017). The scale of VGI in map production: A perspective on European National Mapping Agencies. *Transactions in GIS*, 21(1), 74–90.

Over, M., Schilling, A., Neubauer, S., Zipf, A. (2010). Generating web-based 3D city models from OpenStreetMap: The current situation in Germany. *Computers, Environment and Urban Systems*, 34(6), 496–507.

Palen, L., Soden, R., Anderson, T.J., Barrenechea, M. (2015). Success & scale in a data–producing organization: The socio–technical evolution of OpenStreetMap in response to humanitarian events. In *Proceedings of the 33rd Annual ACM Conference on Human Factors in Computing Systems*, New York.

Perkins, C. (2011). Researching mapping: Methods, modes and moments in the (im)mutability of OpenStreetMap. *Global Media Journal: Australian Edition*, 5(2), 1–12.

Perkins, C. (2014). Plotting practices and politics: (Im)mutable narratives in OpenStreetMap. *Transactions of the Institute of British Geographers*, 39(2), 304–317.

Plantin, J.C. and Valentin, J. (2013). Données ouvertes et cartographie libre. *Les cahiers du numérique*, 9(1), 85–107.

Plennert, M. (2018). The social construction of technological stasis: The stagnating data structure in OpenStreetMap. *Big Data & Society*, 5(2), 1–18.

Plennert, M., Glasze, G., Schlieder, C. (2019). The socio–technical background of an unconventional software architecture in OpenStreetMap: Understanding the implementation of 'Folksonomy'. *Computational Culture*, 7.

Pruvost, H. and Mooney, P. (2017). Exploring data model relations in OpenStreetMap. *Future Internet*, 9(4), 70.

Quest, C. and Bonnet, C. (2016). OpenStreetMap a conduit à la démilitarisation de la carte : interview de Christian Quest, président d'OSM–FR. *Revue internationale d'intelligence économique*, 8(1), 89–92.

Ramm, F., Topf, J., Chilton, S. (2011). *OpenStreetMap: Using and Enhancing the Free Map of the World.* O'Reilly, Sebastopol.

Roche, S., Mericskay, B., Batita, W., Bach, M., Rondeau, M. (2012). Wikigis basic concepts: Web 2.0 for geospatial collaboration. *Future Internet*, 4(1), 265–284.

Sarkar, D. and Anderson, J.T. (2022). Corporate editors in OpenStreetMap: Investigating co-editing patterns. *Transactions in GIS*, 26(4), 1879–1897.

Saunders, A., Scassa, T., Lauriault, T.P. (2012). Legal issues in maps built on third party base layers. *Geomatica*, 66(4), 279–290.

Schröder–Bergen, S., Glasze, G., Michel, B., Dammann, F. (2021). De/colonizing OpenStreetMap? Local mappers, humanitarian and commercial actors and the changing modes of collaborative mapping. *GeoJournal*, 87, 5051–5066.

Senaratne, H., Mobasheri, A., Ali, A.L., Capineri, C., Haklay, M. (2017). A review of volunteered geographic information quality assessment methods. *International Journal of Geographical Information Science*, 31(1), 139–167.

Shulz, S. (2021). De l'adoption au rejet d'un commun numérique pour transformer la frontière entre état et citoyens. *Réseaux*, 225, 151–186.

Singla, N., Sehra, S.S., Singh, J. (2014). A review on vandalism detection in OpenStreetMap data and its emerging trends. *American Journal of Networks and Communications*, 3(6), 77–83.

Sui, D.Z. (2008). The wikification of GIS and its consequences: Or Angelina Jolie's new tattoo and the future of GIS. *Computers, Environment and Urban Systems*, 32, 1–5.

Tang, X., Chong, H.Y., Zhang, W. (2019). Relationship between BIM implementation and performance of OSM projects. *Journal of Management in Engineering*, 35(5).

Tourné, S. (2008). Le travail collectif à distance : enjeux de la communication dans les wikis. PhD Thesis, Université d'Aix–Marseille, Aix en Provence.

Truong, Q.T., Touya, G., Runz, C.D. (2020). Osmwatchman: Learning how to detect vandalized contributions in OSM using a random forest classifier. *ISPRS International Journal of Geo–Information*, 9(9), 504.

Yang, A., Fan, H., Jing, N. (2016). Amateur or professional: Assessing the expertise of major contributors in OpenStreetMap based on contributing behaviors. *ISPRS International Journal of Geo–Information*, 5(2), 21.

Yang, A., Fan, H., Jia, Q., Ma, M., Zhong, Z., Li, J., Jing, N. (2024). How do contributions of organizations impact data inequality in OpenStreetMap? *Computers, Environment and Urban Systems*, 109, 102077.

Zhou, Q. (2018). Exploring the relationship between density and completeness of urban building data in OpenStreetMap for quality estimation. *International Journal of Geographical Information Science*, 32(2), 257–281.

Zhou, Q., Zhang, Y., Ke, C., Brovelli, M.A. (2022). Assessing OSM building completeness for almost 13,000 cities globally. *International Journal of Digital Earth*, 15(1), 2400–2421.

Zielstra, D. and Zipf, A. (2010). A comparative study of proprietary geodata and volunteered geographic information for Germany. In *Proceedings of 13th AGILE International Conference on Geographic Information Science*, Guimarães.

4

Maps and Geographic Information in Planning

Juliette DAVRET
Maynooth University Social Sciences Institute, National University of Ireland, Ireland

4.1. Introduction

In a world increasingly dominated by technology, digitization is transforming every aspect of our lives. Geographic information[1] is becoming essential, both in our daily lives and in political decisions and planning processes. It is essential for planning our public spaces, developing local urban plans, defining housing policy, and marine planning projects. In the field of spatial planning, geographic information serves as the raw material for understanding and analyzing space, often in its most tangible form: maps. It is also the main technical means of implementing development policies to organize space, usually with the help of a plan. Although maps are no longer systematically considered to be a product that stands alone (Gautreau and Noucher 2022), they remain a preferred tool in this field. Geographic information results from the aggregation and organization of data[2] through

1 In this text, information is characterized as *geographic* when it has an explicit relationship with space – that is, when it contains a spatial component that allows it to be mapped in a given geographic reference system.

2 According to Kitchin (2022), data are generally considered as a measurement of the world and the basic element from which information, knowledge and value are produced. It is a measurement collected through various processes that becomes information when it is aggregated, organized and contextualized to produce meaning or understanding of a phenomenon.

processing using technological tools such as geographic information systems (GIS). These tools transform data into a visual and analyzable representation. This facilitates exploring spatial relationships and provides useful knowledge for decision-making.

Spatial planning is an approach that aims to guide the use of space in a territory, taking into account local needs and resources. Unlike other types of planning, like economic or social planning, which focus on specific aspects such as finances or well-being, spatial planning integrates these elements within a geographical perspective. It seeks to organize the use of space for different activities (residential, commercial, industrial, etc.) while preserving the environment. Strategic planning is a component of spatial planning and refers to the set of policies applied to guide the development and activities of an area. Geographic information has become indispensable in planning because it involves a future-oriented approach that requires the development of various scenarios. These allow future actions to be structured in an organized and coherent manner to achieve specific objectives, as planning aims to guide actions over the long term (Fellmann and Morel 1991; Debarbieux and Lardon 2003).

This chapter examines the interdependence and interconnection between geographic information sciences and planning. It aims to show how the use of geographic information and related technologies shapes governance. In addition, it seeks to explain how these elements influence planning practices. To contribute to theoretical debates on planning, this chapter provides an overview of the issues related to the role of geographic information and how it influences planning decisions. Land-based planning has been widely explored through studies on geographic information, often emphasizing inclusive methods involving various actors and collaborative approaches. In contrast, marine planning, which has recently gained momentum in scientific research, tends to adopt a more radically digital-driven approach. This chapter presents a comparison of these two spatial planning processes. By comparing them, the analysis provides a better understanding of how the digital dimension is transforming planning, while highlighting the differences between an inclusive and a more technology-centered approach.

This chapter first explores how digitization has transformed planning practices from a historical perspective (section 4.2). This exploration includes an analysis of theoretical evolution and the growing role of geographic information. The methodological dimension of planning is then discussed and the importance of geographic information in the planning process is examined in depth. This is illustrated through data lifecycles (section 4.3), with a particular focus on marine spatial planning. This leads to a discussion of the sociotechnical network of planning. The relationship between information and participation (section 4.4) is then examined, considering the use of geographic information by and for society, particularly through participatory mapping. Finally, the opportunities and challenges of planning (section 4.5) under the growing influence of the digital turn are discussed, including a critical examination of the democratic illusion it may generate. Key research areas are identified to better understand the challenges of digitization in governance processes. The conclusion emphasizes the implications of these transformations (section 4.6).

4.2. Impact of digitization on planning: a historical perspective

The examination of the role of geographic information and related technologies in planning in scientific literature reveals several trends. The use of digital technologies in data processing and map production dates back to the 1970s. Starting in the 1990s, paper maps began to disappear as a finished product due to the greater accessibility and flexibility of technological tools. The web revolution then transformed map production. Since the 2000s, the emergence of big data and advanced database management systems has enabled the processing and analysis of large volumes of information with greater accuracy and speed. These technologies have fostered the emergence of geographic databases whose visualization in map form is only one facet. Examples include platforms such as Google Maps and OpenStreetMap[3]. The State monopoly on cartography has gradually faded[4]. This has become even more pronounced since the 2010s with the digitization of the world, leading to the "datafication" of spaces and the social relationships built within them (Kitchin and Dodge 2011). This is also transforming the way we think about planning through, for example, the development of smart cities (Cardullo and Kitchin 2019). Finally, digital tools reinforce the growing importance of challenges related to information issues in governance by questioning how stakeholders are involved and decisions are made (Wilson and Tewdwr-Jones 2021). An entire field of research is developing around the digitization of participation tools. This field explores the changes brought about by public participation in data creation, participatory mapping and influencing decision-making through citizen initiatives (Mericskay and Roche 2010).

---

3 Boris Mericskay's contribution (see Chapter 3) details how this iconic tool works.

4 Grégoire Feyt's contribution (see Chapter 5) looks at the ongoing reconfigurations in the production of institutional geographic reference systems.

4.2.1. *Theoretical evolution in planning and the growing role of geographic information*

To understand the changes brought about by geographic information and the digital turn in planning, it is necessary to review the evolution of theoretical models. Without going into detail, it is important to note that comprehensive works exist on the subject and provide much more in-depth coverage of the history of planning. For example, Hartmann and Geertman, in the book by Ansell and Torfing (2016, Chapter 4), discuss planning theories. Dühr (2007) offered an approach to planning focused on spatial visualizations. Therefore, we present here a summary of the historical evolution of this field. In France, the work of Douay (2010, 2013) helps us better understand the evolution of planning approaches. According to his work, planning first adopted a so-called rational model in the 1950s. This model then evolved into a strategic one, in line with the neoliberal economic system of the 1980s. Finally, in the early 2000s, planning shifted to a collaborative model. More recently, the 2010s saw the development of a participatory model. Douay (2013) summarized the main theoretical principles of spatial planning in a multi-entry table. His framework is adapted and expanded here, positioning the role of geographic information (Table 4.1):

	Rational	Strategic	Collaborative	Participatory
Origins	Global rational scientific model (1950)	Neoliberal model (1980)	Interactive model (2000)	Participatory model (2010)
Objectives	Regulating the use of space through science	Achieving results by adopting a pragmatic approach	Interaction between stakeholders to build consensus	Citizen participation to best meet their needs
Key Stakeholders	Policy-makers and planners	Political leaders and economic actors	Political leaders and local stakeholders	Policy-makers, local stakeholders and the general public
Role of geographic information	Technocratic, quantitative and zoning	Quantitative	Based on expert opinions and qualitative trends	Qualitative and consultative

Table 4.1. *Theoretical evolution of planning (adapted from Douay (2013))*

This table illustrates the historical evolution of planning models (rational, strategic, collaborative and participatory). This evolution has also led to changes in the objectives and stakeholders involved in the process, with local communities playing an increasingly important role. Finally, the addition of an entry on the role of geographic information makes it possible to assess the influence of digitization on how planning is carried out. The complexity of the interactions among the different entries in this table calls for an examination of how societies use planning to understand, produce and assign a role to space through geographic information. As Kitchin (2020) pointed out, it is necessary to recognize the codetermined ways in which space is shaped by human action and the digital tools mobilized. It is also crucial to understand how human action influences space and its representation through geographic information and mapping tools. In addition to critical approaches, such as that developed by Kitchin, digital information can be viewed as an intermediary between humans and space in the context of planning. This perspective also fits into other theoretical frameworks, such as the sociology of translation, which offer varied analytical perspectives. To deepen this analysis, it is important to clarify the role of geographic information in the planning process.

4.2.2. *Planning: methods, objectives and geographic technologies*

The objectives and processes for implementing spatial planning can be diverse and must be adapted to the available resources. Nevertheless, it is possible to structure the planning process into five main stages, although these are not exhaustive:

1) territorial assessment, consisting of collecting and analyzing data to understand spatial dynamics;

2) setting goals based on the needs and aspirations of stakeholders;

3) scenarios development to assess the potential impacts of the proposed strategies;

4) implementation of the (spatial) plan, implementing policy choices and regulatory frameworks;

5) monitoring and evaluation to measure progress toward initial objectives.

In summary, spatial planning is an essential tool for forecasting land use while considering the needs of stakeholders. To do this, several components come into play, most of which involve geographic information. Indeed, the use of maps and GIS to visualize and analyze geographic data is almost inherent in every stage of plan implementation. Ehler and Douvere (2009) illustrate this point in their methodological guide for implementing marine planning, from which Trouillet (2019) highlights the role of geographic information (Figure 4.1). As this figure shows, geographic information is essential at every stage, particularly to understand the distribution of resources and activities.

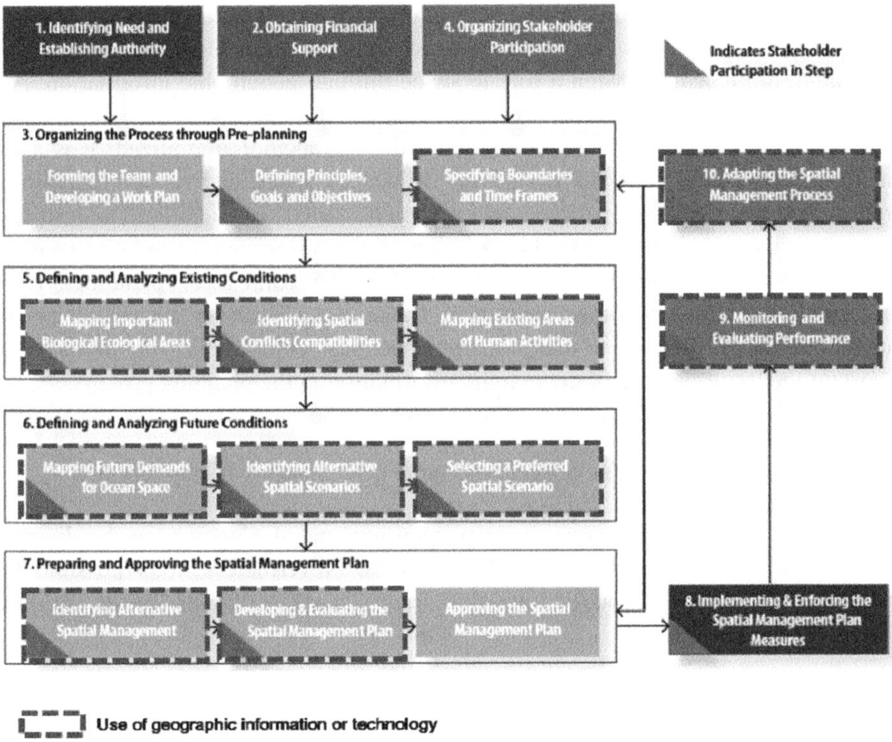

Figure 4.1. *Steps for implementing marine planning (source: adapted from Ehler and Douvere (2009) by Trouillet (2019)). For a color version of this figure, see www.iste.co.uk/joliveau/geovisualization.zip*

However, data digitization tends to erase local differences, reducing the adaptability of datasets to regional specificities in some cases. As Ausello (2024) points out in his examination of urban planning data production in France, there is a tension between two approaches. On the one hand, digitization leads to data standardization, which can result in a technical instrumentalization of territories. On the other hand, there is a reaffirmation of local specificities through more adaptive approaches. This is expressed through the direct involvement of local actors. Balancing these two trends is a dilemma that is particularly apparent in land-based and marine planning, where territorial divisions are often determined by local or regional projects. Therefore, there is a growing contradiction between the need for data to better understand the space to be planned and the difficulty in producing data adapted to local contexts. This is particularly true for marine planning, as some activities at sea remain poorly understood due to a lack of available data (St. Martin and Hall-Arber 2008). To understand how planning makes it possible to anticipate future development actions, it is necessary to examine the lifecycle of geographic information. Planning often relies on incomplete information or data produced, collected and analyzed at a specific point in time. A lifecycle approach helps to understand how these elements influence development decisions.

4.3. Geographic information lifecycle in planning

The geographic information lifecycle plays a key role in spatial governance. It directly influences the quality and relevance of decisions made throughout the planning process. It begins with data collection, essential to ensure the quality and relevance of the knowledge to be deployed. Data are then stored and managed through databases and GIS, facilitating its accessibility and use. Analysis and processing transform the so-called "raw" data into usable information, essential for identifying trends and developing scenarios. Disseminating information to decision-makers and stakeholders through various means (reports, maps, geoportals, etc.) theoretically improves the transparency and effectiveness of governance decisions. Ongoing updates and maintenance of data ensure its relevance and accuracy, enabling decision-making based on the most recent information.

4.3.1. *Application to marine spatial planning*

Marine spatial planning seeks to organize the development of the marine economy while ensuring the achievement of social objectives and maintaining a good environmental status of the seas, based on the "best available scientific knowledge" (Ehler and Douvere 2009). Therefore, it relies heavily on geographic information and related technologies. However, since part of the "reality" is neither captured by existing information nor deployed in the planning process, and given that intrinsic technical limitations of the technologies used persist, the overall role of geographic information and technologies remains largely unexplored.

Our thesis (Davret 2023) examined the potential contributions of a critical analysis of the entire geographic information lifecycle. It focuses on the case of marine planning, broken down into four stages: data production, geographic information processing, representation and finally dissemination. Each stage was examined through a case study, which led to the development of an experiment and a specific methodological approach.

1) Data production was analyzed at the European level through a corpus of 38 marine plans. Examining these plans provides detailed information on the data used in planning and its quality. It revealed that, in the marine context, data production is mainly characterized by the recycling of management data for planning purposes (Davret and Trouillet 2025b). For example, the Wales Marine Plan is based primarily on data provided by the government, most of which was originally collected under the Marine and Coastal Access Act of 2009. These data, initially produced to meet marine stock management objectives, are not specifically suited to the requirements of a marine spatial planning approach. This may pose a practical problem by limiting its relevance for anticipating and organizing interactions between marine uses, thus risking producing decisions that are poorly aligned with the complex challenges characteristics of this space. This mismatch can lead to decisions that, while optimal for a particular sector, neglect global or cross-sectoral challenges, thereby compromising the effectiveness of planning.

2) The processing of geographic information was examined through the establishment of five focus groups that brought together 30 stakeholders (civic society, public institutions and private companies) involved in marine spatial planning in France. Through this participatory method, stakeholders were encouraged to examine the spatial representation of fishing and recreational boating activities (Davret 2024). Analysis of these focus groups led to opening the black box of geomatics processing and identifying stakeholders' knowledge level regarding the processing performed on geographic information (Davret and Trouillet 2025a). Stakeholders acknowledged gaps in the data. For example, recreational boating to date has no spatial tracking at sea for boats less than 20 meters in length. It is therefore very difficult to understand recreational boating practices and map this activity for planning purposes. For fishing, the Vessel Monitoring System (VMS) only covers boats over 12 meters in length, representing around 20% of the French fleet. Consequently, these spatial tracking systems do not counterbalance the existing power dynamics. The lack of comprehensive information available is exacerbated by disparities in understanding geomatics processing. The dominant stakeholders, notably the government, which is the decision-maker, but also fishermen (a key and established stakeholder) or wind farm operators (a powerful new economic stakeholder), also possess an in-depth understanding of mapping tools. This situation accentuates inequalities in the reception and understanding of geographic information.

3) The intelligibility of cartographic representations was examined through five participatory workshops, again bringing together 30 stakeholders. Seven distinct maps, representative of the planning process and produced as part of the French

Marine Spatial Planning documents[5], were examined in detail: strategic maps, environmental and economic inventory maps, maps of a specific marine activity and future-oriented maps. This research analyzed the different elements of the maps and the role each participant assigns to them according to their characteristics to understand the link between their design and their use. This multistakeholder approach, in line with the work of Carton (2007), highlights that the design and use of maps reinforce power relations. These public policy tools used in planning projects contribute to shaping power dynamics among stakeholders and legitimize decisions by marginalizing certain debates. Although they contribute to defining spatial organization, their understanding requires situating them in the context of their production and within the sociopolitical framework of the planning process.

4) Finally, studying the dissemination of geographic information led to the creation of a database of 110 geoportals worldwide, supporting marine planning. This database allows us to examine the features available on these platforms. The participatory format, which is standard practice, creates an illusion of power for marginalized actors (especially civic society). This illusion comes from existing power relationships that strongly influence the decision-making process. In addition, the lack of information tailored to certain marine issues also contributes to this situation. This deficit is attributable to technical and human limitations (Davret et al. 2024).

Geographic information plays a key role in planning, acting both as a decision-making tool and as a means of justifying policy choices. This dual function raises questions about the relationship between the science that produces data, the technical tools that process it and the policy that uses it to guide and legitimize decision-making. Questioning this relationship reveals how domination is structured through the use and understanding of geographic information. Indeed, the activities that can be represented are those for which datasets are available. However, they are also those who have the technical, financial or political means to produce or disseminate spatial representations. In the context of marine planning, this results in economic activities being better represented than cultural activities (such as non-professional fishing, small-scale recreational boating, surfing and other recreational activities). St. Martin and Hall-Arber (2008) highlight this when they examine the "missing layer" of planning. This means that stakeholders with the tools to document their activities have a considerable advantage in defending their interests (e.g. extending their control over certain areas). Consequently, the importance of informational justice in planning – both marine and land-based – extends and reinforces social justice within space. This is why it is essential to

5 The Marine Spatial Plan is a planning document that sets out the guidelines for France's national strategy for the sea and coastline.

approach planning through its sociotechnical aspects, exploring the interplay between knowledge development or technical advancement and social dimensions (Akrich et al. 2006).

4.3.2. *Sociotechnical network in planning*

Analyzing the mechanisms by which geographic information acts on planning and is itself influenced by power-knowledge dynamics in a multistakeholder context enables a better understanding of the complex dynamics of spatial governance. To do this, it is necessary to adopt a methodology centered on the sociotechnical network of geographic information. This methodology must encompass both human components (such as diverse stakeholders) and non-human components (tools or data used). It is also crucial to highlight the close interrelationships between each stage of the information lifecycle. This approach is based on actor-network theory (Latour 2005), which offers a perspective where human and non-human elements coexist and interact inseparably. According to this theory, sociotechnical networks are built through the association of various actors, each with a critical role in information production and circulation. Within this framework, geographic information is not just a set of technical data but also a product of social, political and technological interactions.

Adopting a sociotechnical approach to studying geographic information in a planning context encourages researchers to view planning as an assemblage, according to the concepts of Deleuze and Guattari (1987) taken up by De Landa (2006). The various interpretations of assemblage theory are widely acknowledged as rooted in actor-network theory, given their similarities (Page 2020). An assemblage is a complex system composed of interdependent elements. For example, the planning system combines technical elements (geographic data or technologies) and social elements (human actors or political stakes). This perspective helps to better understand how geographic data are collected, analyzed and circulated in an environment influenced by various actors, where power relations play a key role. This approach also highlights the importance of actors' discourses and practices in the creation and use of geographic information. With digitization and datafication, planning is shifting toward a digital governance model. By extending the concept of assemblage to data (Kitchin 2022; Kitchin and Lauriault 2018), planning becomes the result of multiple data assemblages, which form an emerging and functional data ecosystem (Kitchin et al. 2025a). These data assemblages are interconnected to create a multiscale ecosystem, developing in a contingent, contextual and relational manner through material and discursive practices and processes. The example chosen by Kitchin et al. (2025a) concerns urban planning. It illustrates how the State has replaced paper-based bureaucracy with a set of interconnected data systems, thus forming an ecosystem that is constantly evolving in response to changes in social and technical assemblages.

Although the concept of sociotechnical networks is useful for understanding the digitization and datafication of planning, it remains largely theoretical and needs to be tested through concrete applications to other types of assemblages.

4.4. Information and participation

Geographic information could also play an important role in linking society and planning policies, especially as it could facilitate public participation in the decision-making process. Through the visualization of geographic data, planning projects could become more transparent and accessible, allowing the public to actively engage and contribute to decisions. This participation strengthens the legitimacy of developed plans, as it incorporates the perspectives and needs of a wider community. Geographic information would not merely guide decision-makers; it fosters constructive dialogue between authorities and citizens, promoting more inclusive planning adapted to local realities.

4.4.1. *Understanding the space through geographic information*

Geographic information and related technologies play a key role in participation methods in spatial planning for two reasons. First, because limited physical access to certain areas may require greater reliance on communication technologies and digital tools to increase public participation. For example, in the context of marine planning or protected areas, where direct access is restricted for safety or environmental reasons, tools such as interactive online maps or virtual simulations allow citizens to familiarize themselves with these spaces and contribute to their planning without having to go there physically. Second, because there may be a lack of awareness and understanding of the issues among the public. For example, additional efforts are often needed to educate and raise public awareness of the specific issues of marine spatial planning, for which mapping has become an essential tool. By comparison, people generally have a better understanding and are more aware of land-based issues due to their lived experience. The impacts of land-based planning decisions are often immediate and visible to local communities. In contrast, the direct impact of decision-making in marine planning is less noticeable and more diffuse. This situation highlights the importance of using maps or geoportals to better visualize what is happening beyond the coastal strip. Yet the participatory methods generally used are based on the same formats: public meetings, consultations, surveys, participatory workshops or online platforms.

However, the gap between the participatory objective of planning processes and the active involvement of the public in mapping appears to be widening between land and marine spaces.

Although there is extensive literature on land-based planning on the use of maps to engage the public (Mericskay 2013; McLean et al. 2016), marine examples are scarcer. A few specific cases of participatory mapping for marine space exist, particularly in the context of cultural or recreational[6] activities and marine wildlife observation. The paradox of public participation in marine planning is interesting. On the one hand, it is difficult to engage the public in participatory mapping of a space that few people visit regularly. On the other hand, this participation is an ideal way to learn more about this space and take ownership of it.

While the first three public consultations on marine planning[7] in France merely "skimmed" the maps in the introduction to the planning debate process[8], the consultation entitled the "Sea in Debate"[9] included various mapping exercises. It was a national public debate organized by the National Commission for Public Debate (CNDP). One of the reports published following this consultation (CNDP 2024c) highlights the different uses of mapping. It presents an Atlas of the "Sea in Debate", which brings together maps from government agencies in charge, stakeholders and the public. This initiative brought together participatory mapping methods developed by the Maritime Chair of Nantes University through the "ExperCarto" tool (Giraud et al. 2024) and a simplified mapping portal developed by Cerema[10] to allow web users to draw their own map of the sea online[11]. The ExperCarto tool allowed the public to express their preferred scenarios on the

---

6 The Azores Marine Spatial Plan worked in partnership with the Surfrider Association to create a map of surf spots in order to legitimize coastal protection (see: https://www.surfrider.eu/learn/news/proteger-cote-sauvegardant-vagues-cas-acores/).

7 (1) Preliminary consultation from January 25 to March 25, 2018; (2) public consultation on the first part of the Marine Spatial Planning document from March 4 to June 4, 2019; (3) public consultation on the second part of the MSP document from May 20 to August 20, 2021.

8 Details of the procedures are available at: https://www.merlittoral2030.gouv.fr/.

9 Consultation "Sea in Debate" held between November 2023 and April 2024.

10 Stands for Centre for Studies and Expertise on Risks, the Environment, Mobility and Urban Planning. It is the major French public agency for developing public expertise in the fields of urban planning, regional cohesion and ecological and energy transition for resilient and climate-neutral cities and regions.

11 The online platform is now archived, but more information can be found on the public debate website: https://www.debatpublic.fr/la-mer-en-debat/les-cartes-en-debat-4951#scrollNav-3.

location of offshore wind farms. Participants had to choose from different options, such as clustered or dispersed farms, located near or far from the coast, with or without coexistence of other marine activities. These choices were then translated into simplified spatial representations, generating scenarios or theoretical models of marine development to locate potential wind farms. However, this approach raises questions about the limits of public participation and the technical and environmental constraints that frame these choices. If the tool makes it possible to collect preferred scenarios, it by no means guarantees their implementation, since it is meant to be fictional. Final decisions must consider complex factors such as the protection of marine ecosystems, expert recommendations, and the interests of various stakeholders. If this type of exercise may sometimes prove superfluous given the operational complexity of planning, it is nonetheless a means of actively involving the public. At the same time, the mapping portal developed by Cerema can be used to create maps showing potential areas for marine development, which are then submitted for public debate. These maps can be modified at any time by their authors to incorporate initial feedback from the debate and can be accompanied by additional maps to facilitate discussions in public meetings. However, the flexibility offered by this portal also raises questions about the stability and consistency of proposals, especially when new information or opinions need to be incorporated during the debate.

These two participatory mapping exercises, initially designed to be online, have been adapted for use in face-to-face group workshops, especially during the "Sea in 3D" event and other one-off events in France (Figure 4.2). Public engagement in this process is crucial. However, it must be balanced with a clear understanding of technical and environmental limitations. This ensures that the projects that emerge are both viable and respectful of marine and land ecosystems.

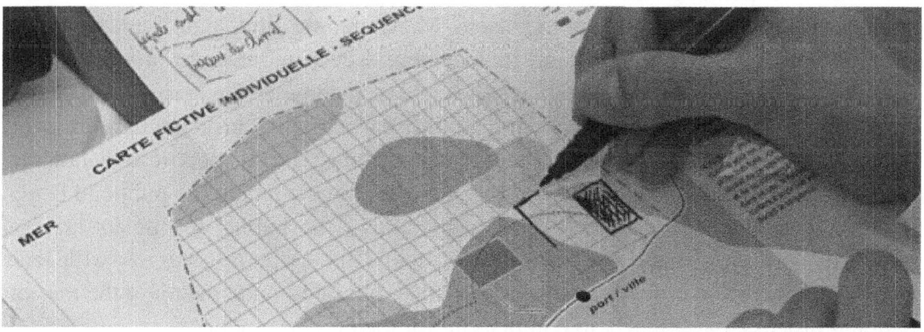

Figure 4.2. *Fictional mapping exercise inspired by the "Expercarto" tool developed by the Maritime Chair of Nantes University (source: CNDP 2024a). For a color version of this figure, see www.iste.co.uk/joliveau/geovisualization.zip*

The CNDP reports repeatedly emphasize the challenge of information and knowledge, particularly because “the sea remains a poorly understood space” (CNDP 2024d, p. 4). The CNDP highlights the importance of the map for public debate, which becomes an educational tool that reveals the complexity of space. Maps “require us to move back and forth between theoretical planning principles and their application to geographical realities. That is, making choices, ranking priorities and identifying their consequences” (CNDP 2024b, p. 3). This latest cycle of public debate on marine planning in France reflects greater awareness of and investment in maps compared to previous debates. It also highlights the reluctance of some members of the public to adopt these tools. The CNDP identifies four main reasons for this refusal:

> Participants believe that it is not their role to draw maps; Participants do not subscribe to the method; Participants believe that they lack the skills required to draw maps; Participants do not wish to locate new offshore wind farms. (CNDP 2024b, p. 36).

Aside from the last reason, which is related to opposition to offshore wind farm development[12], the other categories highlight a lack of legitimacy felt by the public. This lack of legitimacy is particularly obvious during mapping exercises. This highlights a fundamental difference between marine and land-based planning. On land, local stakeholders and the public are regularly consulted and can express their views on issues that directly affect their daily lives, such as road infrastructure or public lighting. By contrast, at sea, consultations generally have less impact on the daily lives of the public. The marine environment is less frequented and therefore less familiar to most people, and the planning challenges are different and can be more complex to understand. In addition, there is a lack of cartographic legitimacy that varies depending on individuals, their socioeducational level and their profession with respect to mapping itself. Many people find it difficult to orient themselves, read a map, understand its legend and even more so to produce maps. This difficulty in mastering mapping tools further exacerbates the perceived distance from the marine environment, making public participation in consultations even more challenging. This distinction highlights the need for a specific and tailored approach to marine planning. This approach must, above all, help the public to better understand marine challenges. It must also involve the public despite its lack of familiarity and daily experience with this space. This situation contrasts with land spaces. It is therefore crucial to better support the public in understanding the marine mapping process, particularly to compensate for the lack of practice.

---

12 The exercise consisted of positioning new offshore wind farms in a fictional space in relation to other activities.

While this tends to evolve, public participation is still limited to the downstream stage of the geographic information lifecycle, which too often excludes data creation and processing, as well as their mapping dimension. As we have shown (Davret 2023), bringing together some decision-makers in the public debate highlights the need to foster collective reflection on the black box effect of marine maps.

4.4.2. *Geographic information and participatory approach*

While many planning and ad hoc decision-making processes are not evidence based, the role of digital technologies and geographic information in participatory approaches tends to foster co-constructed and inclusive decision-making.

Applying the ladder of citizen participation (Arnstein 1969) to planning (Figure 4.3), certain gaps can be observed. Considering the participatory features that could be implemented to involve stakeholders more deeply in decision-making tools, it appears that the highest levels of citizen participation (partnership, delegated power and citizen control) are not fully realized, particularly for marine planning, which seems to face technical limitations.

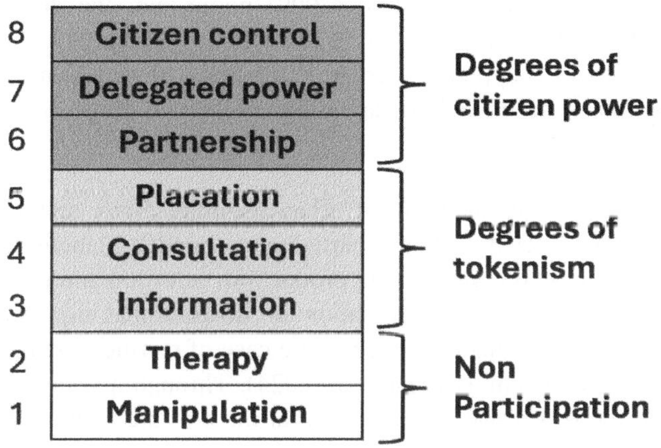

Figure 4.3. *Ladder of citizen participation according to Arnstein (1969)*

As discussed, examining public participation through digital tools and geographic information reveals significant differences between marine and land-based planning. These differences are due to the specific characteristics of the environment and the regulations associated with each type of planning. However, participatory mapping in planning has been tested since at least the 2000s (Joliveau

2001; Kingston 2007; Mericskay and Roche 2010) through participatory GIS tools, participatory geoweb, volunteered geographic information (see OpenStreetMap) and participatory mapping, particularly in future-oriented approaches (Lardon and Noucher 2016). Thus, research has focused primarily on land-based planning and has raised questions about how people understand and use mapping in participatory approaches. Hirt and Roche (2013) published a piece on participatory mapping and explained that there are two relationships or uses linking mapping and participation: in one case, the map is used as support for public debate and participation; in the other case, the public is involved in the mapping itself. The first type of use, which remains the most common, not only disseminates information and creates dialogue, but also manipulates public opinion (Turkucu and Roche 2008). The second allows mapping to be used both as a mediation tool and as the process on which the planning practice is based.

Beyond paper maps, expanding participatory approaches to more complex digital tools, such as GIS, opens up numerous possibilities for the collective management of data and geographic information. However, the technical skills required can limit participation. In addition, the mapping expertise of certain participants may be overestimated, creating a gap between expectations and the reality of their contribution. Nevertheless, as Goodchild (2009) points out, the development of GeoWeb 2.0, which democratizes the production of geographic information, challenges the notion of mapping expertise. Ultimately, this translates into more sporadic contributions from the public than into participation in the overall mapping process.

In their report on the digital future of urban planning, Batty and Yang (2022) highlight a crucial point. Although the participatory model is increasingly being adopted, data-related aspects, such as data production, selection and processing, are not adequately discussed in the consultation or decision-making process. What results from this, as we have demonstrated in the case of marine spatial planning, is a form of participatory illusion (Davret et al. 2024). Through a systematic study of geoportals around the world, we have shown that most of the features available on these platforms are designed primarily for downloading and visualizing data. Thus, these tools help to better understand space. However, very few of the geoportals studied allow active public participation. Even rarer are those that integrate intervention into decision-making. Yet, several authors describe geoportals as supporting marine decision-making (Pınarbaşı et al. 2017; Campbell et al. 2020). This type of geoportal often seems to be merely a facade for planning when made accessible, whereas it could be an active tool for stakeholders and public engagement.

Nevertheless, considering the digital turn in planning, it is worth questioning the theoretical evolution of the participatory model toward a "co-produced" model (Solman et al. 2021). This model is not limited to collaboration between decision-makers and the public, but considers geographic information as a non-human factor capable of actively influencing the planning process. The application of digital technologies to planning (particularly urban planning) is also receiving increasing attention. Geertman and Stillwell (2020) have published a *Handbook of Planning Support Science* that examines how societal challenges can be addressed using digital tools. Other authors also highlight how planning (most studies focus on urban planning) often remains manual, paper-based and tedious (Daniel and Pettit 2021). For example, in the decision-making process for urban infrastructure development and building permits, there is a critical lack of data harmonization. Data are often stored and managed in silos in different organizations, which limits potential collaboration. Moreover, this system requires considerable human and manual work, which can lead to inaccurate, complex and questionable results.

Kayanan et al. (2025) highlight these issues in an article reviewing the Irish planning system. In Ireland, each local authority is responsible for collecting data on planning permission. These data are then sent to the Central Statistics Office for analysis. Although all local authorities receive the same Excel file template to fill out, they return the data in different formats. This is due to human factors, such as manual errors or different interpretations of the instructions. Due to these differences in data formats, the Irish Central Statistics Office then has to spend considerable time reorganizing and harmonizing the data before they can be used, making the process tedious and inefficient. The lack of automation in data collection, from building development application forms to their analysis by the competent authorities, complicates and slows down the process and the use of information. In contrast, in marine planning, digital turn is more significant and radical. However, the use of geographic information technologies remains limited and marginal when it comes to public participation. Indeed, geoportals are mainly accessible after plans have been approved, which prevents the public from having an impact on the decisions made (Davret et al. 2024). This digital turn could be explained by the recent adoption of marine plans that can directly implement digitized systems, whereas urban planning processes have a much more extensive history.

4.5. Digital planning: challenges and opportunities

Digitization is deeply transforming the way spatial planning is conceived and implemented. This phenomenon brings new challenges, particularly in terms of integrating geographic technologies into decision-making and participatory processes. Digital tools make it possible to produce and disseminate data more quickly and interactively, while influencing the dynamics of spatial governance. This shift raises challenges related to the democratic dimension of planning and data management, as well as establishing truly inclusive and participatory governance.

4.5.1. *Rethinking planning by considering geographic information*

Finally, by discussing the different planning models in the digital age, it is possible to reposition marine planning in relation to land-based planning. This can be done, following the framework proposed by Kidd and Ellis (2012), based on the two main areas of debate in spatial planning: on the one hand, the prevailing system of domination, which ranges from expert-based power to a more democratic governance; on the other hand, the epistemological basis of planning, which varies from quantitative to qualitative approaches (Figure 4.4).

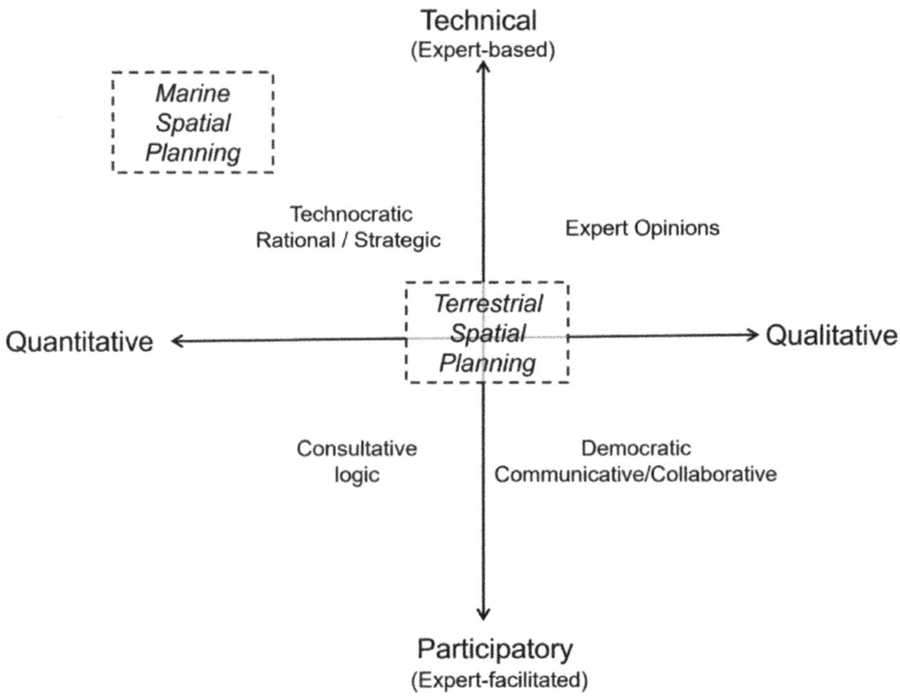

Figure 4.4. *Repositioning planning in relation to geographic information (adapted from Kidd and Ellis 2012)*

While land-based planning strikes a balance between qualitative and quantitative approaches, incorporating technical recommendations from experts and a participatory model involving stakeholders and the general public, marine planning remains largely technocratic. It focuses mainly on quantitative approaches and expert opinions. The participatory approach is limited, and the use of qualitative data has not yet been fully developed.

The parallels between marine and land-based planning highlight the need to reassess the role and influence of geographic information in planning and society. Several paradoxes emerge from this comparison, but one constant remains: the role of maps and geographic information is as essential as it is complex. Land-based planning uses digital tools to support governance, but this digital turn remains fragmented; significant differences between certain services or localities reveal a lack of harmonization. By contrast, marine planning adopts digital tools to support decision-making. These tools take various forms. They are used to produce maps and provide spatial data and information, such as those available on geoportals. However, the approach has not been widely transferred to the public and stakeholders involved in the consultation process. This reinforces the technocratic nature of marine planning compared to land-based planning, which is intended to be more participatory and democratic a priori.

4.5.2. *The digital turn in planning: a democratic illusion?*

While planning is marked by the digital turn, there is an imbalance between the proliferation of digital tool features and non-digital forms of public participation. Tewdwr-Jones and Wilson (2022) show that the technical and procedural approach to digital tools is a barrier to their use by the general public. The authors also note that digital tools aimed at public participation work mainly for immediate issues. The long-term time frame of planning does not correspond to the peak of public attention generated by these features. These same authors also point out that while these tools may attract more people to the planning process, this does not necessarily mean that participation will be more effective. Similar to what we have shown for marine planning (Davret et al. 2024). Technologies and participation processes must be adapted to the way people express themselves, engage and interact.

Some authors criticize the democratic illusion of digital tools (Pritchard and Gabrys 2016; Kloppenburg et al. 2022), particularly because private actors play a key role in providing data and technologies and thereby shape a certain reality. The digital turn in planning could reinforce the imbalance between decision-makers, stakeholders and the public. Even though planning seems to be moving toward a co-production model, it remains primarily a political strategy that reinforces the formalization of knowledge and power relations. This imbalance could be corrected by following the recommendations of Batty and Yang (2022), who emphasize the need not only to "'digitize' the planning system, but fundamentally to 'digitalize' planning methodologies" (p. 10). To do so, the authors recommend making information easily accessible through interactive platforms, such as online geoportals, which allow citizens to consult urban or territorial plans and track their development. They also advocate for real-time monitoring of plans via digital dashboards, like the project management applications used in some cities to monitor various urban indicators (Kitchin et al. 2015). In addition, to better involve stakeholders, online consultation tools and participatory surveys could be set up, such as platforms like "Decidim"[13] in Barcelona, which allowed citizens to submit their ideas and actively participate in the co-construction of the city's municipal action plan for the period 2016–2019. Improving transparency and efficiency is crucial to ensure the traceability and security of data exchanged between planning authorities, private companies and communities. Finally, it is essential to ensure that the process does not become purely technocratic in justifying digitization. Initiatives such as smart cities, where technology is not only used for technical management but also to strengthen collaboration between various actors while keeping citizens at the heart of decisions, should be used as examples (Deng et al. 2021). According to Potts (2020), this requires connecting planning theories to the impact of digitization on our society.

By establishing the principles necessary for a digital approach to multiscale, multistakeholder planning, a new step forward can be taken. This entails placing greater importance on data and improving its updating and diversity. According to Batty and Yang (2022), these changes could usher in a new era of planning. The authors emphasize that the digitization of planning would strengthen stakeholders and public engagement, and break down silos between different decision-making bodies. The integration of digital tools into spatial planning, promoted by theorists such as Batty, paves the way for sophisticated but inherently complex systematic modeling. This approach offers potentially powerful solutions for anticipating and managing spatial dynamics, but it also presents a significant risk: that of the decision-making process being hijacked by an elite group of experts. By relying on algorithmic logic that is often unintelligible to non-experts, this digital and quantitative bias can create a divide between decision-makers or technical experts and the general public. Citizen participation, already fragile, could be pushed to the background on the pretext that local populations lack the legitimacy to intervene in planning processes deemed too complex to be understood by laypeople, across the three levels highlighted by this chapter: mapping, geomatics and modeling. This

13 See: https://decidim.org/fr/.

raises a major contradiction: How can we reconcile the need for inclusive local governance with tools whose technical sophistication de facto excludes non-experts?

To avoid this drift, it is essential to develop mechanisms that make these models more transparent and accessible, while revaluing local knowledge and citizen perspectives in decision-making. Thus, participatory geoprospective emerges as a promising solution (Gourmelon et al. 2012; Houet and Gourmelon 2014). It is based on an exploratory approach that aims to construct possible scenarios using geographic information to inform public action. As a future-oriented approach (Godet 1986), it considers possible developments in a territory with regard to public action (Lardon and Noucher 2016). Participatory geoprospective can therefore be seen as "the informational side of geogovernance" (Lardon and Noucher 2016, p. 212). By making possible scenarios visible, it challenges the nature of the information that feeds into computer systems and modeling. The study conducted by Dodane et al. (2014) on the Yzeron watershed (France) shows that geoprospective helps structure debates and fosters collective engagement, which generates exchanges between the different stakeholders on future scenarios. Participatory geoprospective enables participants to better identify areas likely to change and, above all, assess the effects of planning. Nevertheless, the participants in this experiment point out that not all elements to be considered are included in the simulation, particularly qualitative aspects that cannot be mapped. The authors, meanwhile, highlight the participants' lack of critical thinking regarding the proposed models concerning the past. Emsellem et al. (2012) point out another important limitation: integrating actors into geoprospective is difficult due to their great diversity and the complexity of the roles they play in a territory. This inclusion requires long term commitment, which contrasts with the short time often allocated to the decision-making process. This implies rethinking the place of participation in a context where technical complexity and lack of time should not become pretexts for democratic exclusion. Thus, public participation through digital tools must be considered a genuine methodological stance in the decision-making process. It must be detached from post-political logic, as is currently the case in marine planning. The digital turn in planning would no longer simply be a "governance tool" to illustrate the actions of decision-makers; it would also become a "mode of governance" to engage stakeholders and the public in the process and make more democratic and inclusive decisions (Campbell et al. 2020, p. 296).

Nevertheless, fully and unreservedly digitizing planning is probably not the solution. It is crucial to adopt a critical approach to geographic information and maps. Such a perspective highlights the complexities, ambiguities and power dynamics involved in the production and use of geographic information. To this end, two major research areas must be undertaken in the coming years.

4.5.3. *Research agenda*

Unpacking the role of geographic information in planning involves examining how that information is produced, analyzed and disseminated across the infrastructure that generates or uses spatial data and the stakeholders who rely on it. The field of critical data studies offers an analytical framework that encompasses these elements to reconsider sociotechnical networks of planning and data mobility. The latter has been described and conceptualized in various ways: for example, data mobility is referred to as flows, streams, journeys, threads, exchanges, circulation and transfers in scientific literature. As Kitchin et al. (2025b) point out, data mobility is often, if not always, conceived as a movement from one point to another. This model visualizes data moving along a chain between producers, senders and receivers. However, this does not reflect the digital circulation of data and geographic information. Data is multiplied and proliferated. It is copied from one system to another and, in the process, transformed by merging, cleaning or manipulating. Consequently, this theoretical approach to planning digitization should help us rethink the movement of data and its interactions within a sociotechnical ecosystem where different narratives, actors and systems coexist and evolve together.

From an empirical perspective, it is necessary to strengthen the documentation of the sociotechnical entanglements of planning by carefully tracing the path of data from its production to its dissemination. Deepening our knowledge and understanding of data trajectories throughout their lifecycle makes it possible to better grasp the processes that shape and transform it into different contexts. This leads to a deeper understanding of the data ecosystem, illustrated through examples, and ultimately to a more effective governance of digital planning systems.

4.6. Conclusion

This chapter attempts to summarize the challenges of digitization in spatial planning. It has shown the importance of questioning the role of digitization in planning processes and its implications for power relations.

One of the main challenges emerging from the comparison between land-based and marine planning is the role of participation in relation to mapping tools and media. There is a notable difference in terms of accessibility, the involvement of a wide range of stakeholders and the level of knowledge and awareness among communities. Marine planning often requires tailored strategies to effectively engage the public. Digital tools seem to offer a suitable response but remain underdeveloped to fully meet this need. Conversely, in land-based planning, local communities tend to participate actively, yet local systems themselves sometimes appear disconnected from the digital turn. While, in theory, attention is turning to a new co-produced planning system, in practice, the flow of data between public consultation and decision-making remains largely unclear.

Geographic information plays a dual role in planning: it both legitimizes and shapes decisions. This results in a representation of issues that is partial, and sometimes biased, a reality that is continually challenging in the field of critical cartography and, more broadly, critical data studies. Analyzing the lifecycle of geographic information in planning helps reveal the biases and misuses inherent at each stage, which tend to accumulate and intensify from data production to its circulation. Information-related challenges are distributed throughout the planning process, rather than being concentrated at the early stage of data creation. It is therefore essential to make the informational dynamic explicit in planning within the digital age and to contribute to the ongoing critical shift currently underway in this field.

Accordingly, in line with the research agenda focusing on the role of geographic information in planning, it is important to shed light on the complexities and power dynamics involved in its production and use. It is also essential to explore approaches that promote more inclusive and critical planning practices, an area of research that warrants further exploration to deepen reflection and generate concrete solutions.

4.7. References

Akrich, M., Callon, M., Latour, B. (eds) (2006). *Sociologie de la traduction. Textes fondateurs*. Presses des Mines, Paris. doi: 10.4000/books.pressesmines.1181.

Ansell, C. and Torfing, J. (eds) (2016). *Handbook on Theories of Governance*. Edward Elgar Publishing, Cheltenham. doi: 10.4337/9781782548508.

Arnstein, S.R. (1969). A ladder of citizen participation. *Journal of the American Institute of Planners*, 35(4), 216–224. doi: 10.1080/01944366908977225.

Ausello, N. (2024). La fabrique publique/privée des données de planification urbaine en France : entre logique gestionnaire et approche territorialisée de la règle. *Cybergeo: European Journal of Geography*. doi: 10.4000/11o4n.

Batty, M. and Yang, W. (2022). A digital future for planning. Spatial planning reimagined. Digital Task Force for Planning [Online]. Available at: https://digital4planning.com/wp-content/uploads/2022/02/A-Digital-Future-for-Planning-Full-Report-Web.pdf.

Campbell, L.M., St. Martin, K., Fairbanks, L., Boucquey, N., Wise, S. (2020). The portal is the plan: Governing US oceans in regional assemblages. *Maritime Studies*, 19(3), 285–297. doi: 10.1007/s40152-020-00173-3.

Cardullo, P. and Kitchin, R. (2019). Smart urbanism and smart citizenship: The neoliberal logic of 'citizen-focused' smart cities in Europe. *Environment and Planning C: Politics and Space*, 37(5), 813–830. doi: 10.1177/0263774X18806508.

Carton, L.J. (2007). Map making and map use in a multi-actor context: Spatial visualizations and frame conflicts in regional policymaking in the Netherlands. PhD Thesis, Radboud University [Online]. Available at: https://repository.tudelft.nl/record/uuid:f5821d4b-5299-4af9-baad-176dbf1daabb.

CNDP (2024a). La mer en débat – Environnement, éolien en mer, activités maritimes et littoral. Commission Nationale du Débat Public [Online]. Available at: https://www.debatpublic.fr/sites/default/files/2024-05/Synthese-courte-lameren3D_0.pdf.

CNDP (2024b). Atlas "La mer en débat" 20 novembre 2023-26 avril 2024. Commission Nationale du Débat Public [Online]. Available at: https://www.debatpublic.fr/sites/default/files/2024-06/DSF-atlas-la-mer-en-debat.pdf.

CNDP (2024c). Compte rendu "La mer en débat" 20 novembre 2023-26 avril 2024. Commission Nationale du Débat Public. [Online]. Available at: https://www.debatpublic.fr/sites/default/files/2024-06/DSF-Compte-rendu.pdf.

CNDP (2024d). Synthèse du compte rendu du débat public. 20 novembre 2023-26 avril 2024. Commission Nationale du Débat Public [Online]. Available at: https://www.debatpublic.fr/sites/default/files/2024-06/DSF-synthese-compte-rendu.pdf.

Daniel, C. and Pettit, C. (2021). Digital disruption and planning – Use of data and digital technology by professional planners, and perceptions of change to planning work. *Australian Planner*, 57(1), 50–64. doi: 10.1080/07293682.2021.1920995.

Davret, J. (2023). La partie immergée de l'information géographique : analyse critique à travers le cas de la planification maritime. PhD Thesis, Nantes Université [Online]. Available at: https://theses.hal.science/tel-04123271.

Davret, J. (2024). Déconstruire les cartes maritimes : révéler les choix cartographiques à travers la pêche et la plaisance. *L'Atlas Bleu, Revue cartographique des mers et des littoraux* [Online]. Available at: https://atlas-bleu.cnrs.fr/deconstruire-les-cartes-reveler-les-choix-cartographiques/.

Davret, J. and Trouillet, B. (2025a). How do stakeholders engage with critical cartography in planning? Analysis of a decision-making process in marine governance. *Environmental Science & Policy*, 169, 104083. doi: 10.1016/j.envsci.2025.104083.

Davret, J. and Trouillet, B. (2025b). Recyclage des données pour la planification spatiale marine : examen des plans maritimes en Europe. *Journal of Interdisciplinary Methodologies and Issues in Sciences*, 12. doi: 10.46298/jimis.13496.

Davret, J., Trouillet, B., Toonen, H. (2024). The digital turn of marine planning: A global analysis of ocean geoportals, *Journal of Environmental Policy & Planning*, 26(1), 75–90. doi: 10.1080/1523908X.2023.2283081.

De Landa, M. (2006). *A New Philosophy of Society: Assemblage Theory and Social Complexity*. Continuum, London, New York.

Debarbieux, B. and Lardon, S. (2003). *Les figures du projet territorial*. Éditions de l'Aube, Datar [Online]. Available at: https://shs.hal.science/halshs-00280078.

Deleuze, G. and Guattari, F. (1987). *A Thousand Plateaus: Capitalism and Schizophrenia*. University of Minnesota Press, Minneapolis.

Deng, T., Zhang, K., Shen, Z.-J. (2021). A systematic review of a digital twin city: A new pattern of urban governance toward smart cities. *Journal of Management Science and Engineering*, 6(2), 125–134. doi: 10.1016/j.jmse.2021.03.003.

Dodane, C., Joliveau, T., Rivière-Honegger, A. (2014). Simuler les évolutions de l'utilisation du sol pour anticiper le futur d'un territoire. *Cybergeo: European Journal of Geography*. doi: 10.4000/cybergeo.26483.

Douay, N. (2010). Collaborative planning and the challenge of urbanization: Issues, actors and strategies in Marseilles and Montreal metropolitan areas. *Canadian Journal of Urban Research*, 19, 50–69.

Douay, N. (2013). La planification urbaine française : théories, normes juridiques et défis pour la pratique, *L'Information géographique*, 77(3), 45–70. doi: 10.3917/lig.773.0045.

Dühr, S. (2007). *The Visual Language of Spatial Planning: Exploring Cartographic Representations for Spatial Planning in Europe*, 1st edition. Routledge, London. doi: 10.4324/9780203965818.

Ehler, C. and Douvere, F. (2009). *Marine Spatial Planning: A Step-by-Step Approach toward Ecosystem-based Management*. UNESCO, Paris [Online]. Available at: https://unesdoc.unesco.org/ark:/48223/pf0000186559.

Emsellem, K., Liziard, S., Scarella, F. (2012). La géoprospective : l'émergence d'un nouveau champ de recherche ? *L'Espace géographique*, 41(2), 154–168. doi: 10.3917/eg.412.0154.

Fellmann, T. and Morel, B. (1991). Territoires en action. Prospective urbaine et planification stratégique. *Les Annales de la Recherche Urbaine*, 51(1), 93–102. doi: 10.3406/aru.1991.1603.

Gautreau, P. and Noucher, M. (2022). Farewell to maps: Reformulating critical cartography in the digital age. In *The Politics of Mapping*, Debarbieux, B. and Hirt, I. (eds). ISTE Ltd, London and Wiley, Hoboken, NJ.

Geertman, S. and Stillwell, J. (2020). *Handbook of Planning Support Science*. Edward Elgar Publishing, Cheltenham.

Giraud, A., Pigeault, A., Boilevin, V., Rollo, N., Mahieu., P-A., Wolff, F-C., Trouillet B. (2024). Synthèse de l'analyse des contributions de l'étude "EXPERCARTO : les préférences des français sur l'éolien en mer "Nantes Université [Online]. Available at: https://hal.science/hal-04635144v1.

Godet, M. (1986). Introduction to la prospective. *Futures*, 18(2), 134–157. doi: 10.1016/0016-3287(86)90094-7.

Goodchild, M. (2009). NeoGeography and the nature of geographic expertise. *Journal of Location Based Services*, 3(2), 82–96. doi: 10.1080/17489720902950374.

Gourmelon, F., Houet, T., Voiron-Canicio, C., Joliveau, T. (2012). La géoprospective, apport des approches spatiales à la prospective. *L'Espace géographique*, 41(2), 97–98. doi: 10.3917/eg.412.0097.

Hartmann, T. and Geertman, S. (2016). Planning theory. In *Handbook on Theories of Governance*, Ansell, C. and Torfing, J. (eds). Edward Elgar Publishing, Cheltenham. doi: 10.4337/9781782548508.00013.

Hirt, I. and Roche, S. (2013). Cartographie participative. In *Dictionnaire critique et interdisciplinaire de la participation*, Casillo, I., Barbier, R., Blondiaux, L., Chateauraynaud, F., Fourniau, J-M., Lefebvre, R., Neveu, C., Salles, D. (eds). GIS Démocratie et Participation, Paris [Online]. Available at: http://www.dicopart.fr/fr/dico/cartographie-participative.

Houet, T. and Gourmelon, F. (2014). La géoprospective – Apport de la dimension spatiale aux démarches prospectives. *Cybergeo*. doi: 10.4000/cybergeo.26194.

Joliveau, T. (2001). La participation à la décision territoriale : dimension sociogéographique et enjeux informationnels d'une question politique. *Géocarrefour* 76, 273–279.

Kayanan, C.M., Mutter, S., Davret, J., Kitchin, R. (2025). The digital turn in planning and the production of 'good enough' planning systems. *European Planning Studies*, 1–22. doi: 10.1080/09654313.2025.2599869.

Kidd, S. and Ellis, G. (2012). From the land to sea and back again? Using terrestrial planning to understand the process of marine spatial planning. *Journal of Environmental Policy & Planning*, 14(1), 49–66. doi: 10.1080/1523908X.2012.662382.

Kingston, R. (2007). Public participation in local policy decision-making: The role of web-based mapping. *The Cartographic Journal*, 44(2), 138–144. doi: 10.1179/000870407X213459.

Kitchin, R. (2020). Space and spatiality. In *International Encyclopedia of Human Geography (Second Edition)*, Kobayashi, A. (ed.). Elsevier, Oxford. doi: 10.1016/B978-0-08-102295-5.10998-9.

Kitchin, R. (2022). *The Data Revolution: A Critical Analysis of Big Data, Open Data & Data Infrastructures*, 2nd edition. Sage Publications, Los Angeles, CA.

Kitchin, R. and Dodge, M. (2011). *Code/Space: Software and Everyday Life*. MIT Press, Cambridge, MA.

Kitchin, R. and Lauriault, T.P. (2018). Towards critical data studies: Charting and unpacking data assemblages and their work. In *Thinking Big Data in Geography*, Thatcher, J., Eckert, J., Shears, A. (eds). University of Nebraska Press, Lincoln.

Kitchin, R., Lauriault, T.P., McArdle, G. (2015). Knowing and governing cities through urban indicators, city benchmarking and real-time dashboards. *Regional Studies, Regional Science*, 2(1), 6–28. doi: 10.1080/21681376.2014.983149.

Kitchin, R., Davret, J., Kayanan, C.M., Mutter, S. (2025a). Assemblage theory, data systems and data ecosystems: The data assemblages of the Irish planning system. *Big Data & Society*, 12(3), 20539517251352822. doi: 10.1177/20539517251352822.

Kitchin, R., Davret, J., Kayanan, C.M., Mutter, S. (2025b). Data mobilities: Rethinking the movement and circulation of digital data. *Mobilities*, 20(5), 1002–1020. doi: 10.1080/17450101.2025.2481309.

Kloppenburg, S., Gupta, A., Kruk, S.R.L., Makris, S., Bergsvik, R., Korenhof, P., Solman, H., Toonen H.M. (2022). Scrutinizing environmental governance in a digital age: New ways of seeing, participating, and intervening. *One Earth*, 5(3), 232–241. doi: 10.1016/j.oneear.2022.02.004.

Lardon, S. and Noucher, M. (2016). Construire demain par les cartes : usages de l'information géographique en prospective territoriale participative. *Cahiers de géographie du Québec*, 60(170), 209–219. doi: 10.7202/1040531ar.

Latour, B. (2005). *Reassembling the Social: An Introduction to Actor-Network-Theory*. Oxford University Press, Oxford, New York.

McLean, J., Maalsen, S., Grech, A. (2016). Learning about feminism in digital spaces: Online methodologies and participatory mapping. *Australian Geographer*, 47(2), 157–177. doi: 10.1080/00049182.2016.1138842.

Mericskay, B. (2013). Cartographie en ligne et planification participative. Analyse des usages du géoweb et d'Internet dans le débat public à travers le cas de la Ville de Québec. PhD Thesis, Université de Laval [Online]. Available at: https://www.collectionscanada.ca/obj/thesescanada/vol2/QQLA/TC-QQLA-29535.pdf.

Mericskay, B. and Roche, S. (2010). Cartographie numérique en ligne nouvelle génération : impacts de la néogéographie et de l'information géographique volontaire sur la gestion urbaine participative. In *HyperUrbain II, nouvelles cartographies, nouvelles villes*, Zreik, K. (ed.). Europia, Paris [Online]. Available at: https://shs.hal.science/halshs-02309592.

Page, S. (2020). Assemblage theory. In *International Encyclopedia of Human Geography (Second Edition)*, Kobayashi, A. (ed.). Elsevier, Oxford. doi: 10.1016/B978-0-08-102295-5.10506-2.

Pınarbaşı, K., Galparsoro, I., Borja, Á., Stelzenmüller, V., Ehler, C.N., Gimpel, A. (2017). Decision support tools in marine spatial planning: Present applications, gaps and future perspectives. *Marine Policy*, 83, 83–91. doi: 10.1016/j.marpol.2017.05.031.

Potts, R. (2020). Is a new "planning 3.0" paradigm emerging? Exploring the relationship between digital technologies and planning theory and practice. *Planning Theory & Practice*, 21(2), 272–289. doi: 10.1080/14649357.2020.1748699.

Pritchard, H. and Gabrys, J. (2016). From citizen sensing to collective monitoring: Working through the perceptive and affective problematics of environmental pollution. *GeoHumanities*, 2(2), 354–371. doi: 10.1080/2373566X.2016.1234355.

Solman, H., Smits, M., van Vliet, B., Bush, S. (2021). Co-production in the wind energy sector: A systematic literature review of public engagement beyond invited stakeholder participation. *Energy Research & Social Science*, 72, 101876. doi: 10.1016/j.erss.2020.101876.

St. Martin, K. and Hall-Arber, M. (2008). The missing layer: Geo-technologies, communities, and implications for marine spatial planning. *Marine Policy*, 32(5), 779–786. doi: 10.1016/j.marpol.2008.03.015.

Tewdwr-Jones, M. and Wilson, A. (2022). Co-designing urban planning engagement and innovation: Using LEGO® to facilitate collaboration, participation and ideas. *Urban Planning*, 7(2). doi: 10.17645/up.v7i2.4960.

Trouillet, B. (2019). Aligning with dominant interests: The role played by geo-technologies in the place given to fisheries in marine spatial planning, *Geoforum*, 107, 54–65. doi: 10.1016/j.geoforum.2019.10.012.

Turkucu, A. and Roche, S. (2008). Classification fonctionnelle des public participation GIS. *Revue internationale de géomatique*, 18(4), 429–442. doi: 10.3166/geo.18.429-442.

Wilson, A. and Tewdwr-Jones, M. (2021). *Digital Participatory Planning: Citizen Engagement, Democracy, and Design*. Routledge, New York.

5

GIS Dissemination Facing Territorial Disparity

Grégoire FEYT
UMR PACTE, Université Grenoble Alpes, France

"We know the sound of two hands clapping,
but what is the sound of one hand clapping?"
Zen proverb quoted by J.D. Salinger in Short Stories (1953)

5.1. Introduction

At first glance, we would expect that digital geographic information is now available everywhere, for everyone, and on most things. The widespread use of geolocation, mobile technologies, sensors, etc., combined with the growing availability of aerial photographs, base maps, generic databases and associated processing tools, shape our individual and collective practices, and the way our territories develop.

However, geographic information as a description of "real space" has the peculiarity of being fractal, so to speak. There is no better way to describe this than to refer to Borges' famous and very short story about the *Map of the Empire* (Joliveau 2007), in which cartographers, dissatisfied with having a map of the Empire that was "only" the size of a province, decide to embark on the creation of a 1:1 map, thereby signing the end of cartography. We can thus conceive the process of geonumerical description of the "real world" in two symmetrical ways: one "top-down", starting from a global but summary understanding that is gradually refined; the other, "incremental", based on detailed descriptions of parts that gradually complement and aggregate to make a representation of the whole. Both approaches do coexist while belonging to fundamentally different professional spheres, practices and rationales. The first, more recent, mainly arises from the combined action of the market and technological developments. The second relates to the professions of land management and planning; it is also part of a "path dependency" that is largely dependent on institutional and organizational factors, as on geographical factors. While digital geographic information methods and tools were initially the preserve of what could be called "territorial geomatics", they have now been diluted into a variety of actual or potential fields of application, services and knowledge production. Their eclectic and "disruptive" nature tends to attract media and even scientific attention.

This territorial geomatics project is still ongoing and can be approached twofold: one performative with a focus on how GIS, geomatics resources, new tools, new data, new fields of application, etc., can help territories with these capabilities; the other is strictly territorial, rooting from the diverse territories and seeking how they are able to use these new and effective means of producing self-knowledge. In terms of research, there is a rich and diverse literature on "GIS and [such-and-such a field]" or "geomatics in the service of [...]". Attempting to list the actual or potential resources of GIS would be a necessarily incomplete and reductive endeavor. However, according to scientific literature published in recent years, the production of geospatial resources by and for local authorities does not seem to be a theoretical issue any longer. While this was not always the case, the existence of these resources now seems taken for granted, one element of context among others and no longer a marker revealing wider problematic or issue.

The golden age of local GIS development and the dematerialization of geographic information began in France in the late 1990s. It gave rise to fifteen years of scientific research and studies aimed at understanding how these new tools, designed to inform and implement an integrated and operational approach to various areas of territorial public action, could find their place and legitimacy within technostructures organized in silos and based on distinct and often disjointed powers and skills. In this context and from this perspective, the contributions of research have been invaluable and there is no need to revisit, in detail, work corresponding to a now outdated or overcome situation. However, the multidimensional lessons learned from this research allow us to understand how the technical, functional, organizational and even institutional constraints and expectations of implementing local GIS had to contend with the territory conceived as a complex system (Moine 2006). With that in mind, the territory appears indeed as the product of a permanent dynamics of interactions between natural space, inhabited, exploited and shaped geographical space, but also organized or unorganized actors. The basic premise, we might say, of the "datafication" of the territory has been clearly demonstrated (Major 1999) by showing, through lexical analysis, that practitioners of territorial decision-making and action (urban planning, transportation, heritage, environment, etc.) construct and mobilize very different cognitive models for the same space, according to their respective professions. The consequence, when it comes to materializing this cognitive model in the form of a data model that can be used by a computer, has been expressed clearly and concisely by Noucher (2012): "geographic data is the construction of a territorial reality that reflects the experience and expectations of its producer". Under these conditions, far from being a smooth process, the development and legitimization of a GIS within a territorial structure needs to coordinate strongly sectoral visions of the territory. As such, it becomes, as Henri Pornon observed in the late 1990s (Pornon 1998), an object of power struggles and therefore of negotiation, reflecting an undeniable reality that must be dealt with: "at the territorial level, as within the organization responsible for its management and future, there is no overall rationale that everyone can agree on". Thus, GIS, *between organization and territory* (Joliveau 2004), are the result of a variable arbitration between operational requirements and the desire to contribute to an integrated knowledge of the territory that can be mobilized in the context of debate and collective decision-making. However, often dominated by a combination of urban and computer engineering, the creation of GIS within local authorities has often tended to favor the operational and technical aspects. In 2002, Marie-Hélène de Sède-Marceau, quoted by Alexandre Moine (Moine 2006), considered GIS to be "accumulations of data" that did not (yet) allow for the design of real territorial observatories integrating the interplay of actors and the forces that drive it through appropriate descriptors. At the same time, however, as Noucher (2012) points out, unlike other computer applications such as accounting software, geographic information technologies do not offer users a closed environment delimited by a finite set of functionalities and data. Geographic data thus act as a gateway, opening up other services and enabling other uses and applications. Hence, once geospatial resources have found their place within the structure, proven their usefulness and ensured their visibility, the scope of their uses and possibilities tends to expand and diversify naturally, boosted by the continuous emergence of new data and scientific or technical advances. This is clearly the type of dynamics at work today, yet mostly limited to large urban communities, as we will see later.

This last paragraph, which is necessarily simplistic, aims to highlight the specific difficulties faced by local authorities that began developing GIS systems, mainly from the 2000s onward. We will return to the more factual and structural dimensions of what was ultimately a fairly long process when compared to the emergence and widespread use of digital technology applied to other types of information such as text, images, sound, music notation, etc. These elements are needed to understand the factors that have governed and continue to govern the dissemination – and the non-dissemination – of GIS within territories. In any case, the issue of implementing GIS within regional organizations has gradually ceased to be of real scientific interest, with the collective work published in 2009 entitled *Organizational Aspects of GIS* (Roche and Caron 2009). In a way, this collaborative publication validated the fact that choices had been made, situations had stabilized and that it was now time to move more broadly into the "geographic information society".

Therefore, while we can consider that GIS no longer have a problem with territory, we can still ask ourselves whether territories have a problem with GIS. The situation is different if, instead of detailing "what geomatics brings or can bring to territories", we look at it from the perspective of the territories themselves, questioning their ability to acquire the material, human, cognitive and organizational resources necessary to produce and leverage specific data that can feed into self-knowledge. In doing so, we approach the issue of geographic information from both a scientific and political perspective, but through the same concept of territorial continuity of knowledge. Indeed, partly for historical reasons, cartographic representation in France and its perpetuation through digital geographic information is inseparable from the concept of public action, which is itself inseparable from the principle of territorial continuity of public action, with the implicit corollary of the necessary continuity of territorial knowledge itself. The term *territory* is used here in a very specific sense to refer to the concept of territorial institutions. This clarification is necessary as the concept of territory has been and continues to be the subject of a wealth of literature exploring its many dimensions and manifestations; since 2010, this work has even found a unifying framework with the establishment of *territorial sciences* (Beckouche 2012). However, given the subject at hand, we believe it is not useful to extend the scope of our reflection to concepts such as territoriality or governance issues: GIS have been and still are, to a large extent, tools produced by and for professionals within "technical–political" structures with explicit powers within a defined space using allocated resources. The question is now to determine to what extent and under which conditions are very diverse territories able to increase their "self-knowledge" with the use of current and future geodata, methods and tools. When not able, then a scientific and political issue arises: What other forms of knowledge and knowledge production can be devised to avoid falling through the net, knowing, as we shall see, that in this case the net is more loose than tight?

Our goal consists not only in identifying the "territorial landscape" of GIS in France today, which we will see is highly contrasting, but also of understanding the factors that have led to this situation. Unfortunately, we do not have any recent surveys or publications to draw on, hence this chapter being more of an investigative approach than a state-of-the-art review. Beyond the empirical observations that have guided and fed our thought, we sought to find "collateral" data that could yield meaningful findings. We also relied heavily on government and socioprofessional reports, whose very existence often reveals an awareness of "real-world" problems and/or a desire to influence developments. In a way of speaking, we sought to produce an "X-ray of GIS in the territories" using a set of converging indicators. The resulting diagnosis will lead us to propose research directions encouraging the close integration of territorial sciences and geographic information sciences.

5.2. From national puzzle to urban patchwork

The territory is also THE Territory, in the sense of the national territory. This is especially true both in the French case and when dealing with geographic information. For a long time, map production was essentially the monopoly of the state through its ministries and specialized institutions, foremost among which were the Geographic National Institute (Institut Géographique National (IGN)) and the Tax Directorate (Direction Générale des Impôts (DGI)) in charge of the cadastre management. With its powers and skills, the state took up the issue of dematerializing what were then known as *plans* and *maps* (and not yet geographic data) (Feyt 2002). The French state thus focused on two categories of information that are still integral to the role and even the very existence of the state today:

– the land registry for the purpose of guaranteeing land ownership and taxation;

– base maps that enable the national territory to be described by providing a spatially continuous reference system that is unique and common to both semantic and semiological plans.

5.2.1. *Major public works projects*

With the growth of its digital heritage of territorial data, and at the same time bolstered by a prevailing belief that cybernetics would enable cognitive and organizational barriers to be broken down, the state began taking steps in the late 1990s to implement cross-functional systems and tools for understanding the territory based on principles it had predefined. One example is the project launched in 1999 to provide each French department or region with a *Territorial Information System* (Système d'Information Territorial (SIT)) shared by government departments, according to an implementation guide. In November 2000, a report by the *Government Mission on Information and Communication Technologies* noted with satisfaction that "cooperative working protocols are in the process of being standardized" and that the systems put in place "also aim to develop and manage individual and collective skills". There was talk of "knowledge management tools" when the objective was "simply" to collect and consolidate data. In fact, while it is nowadays possible to find a number of references to departmental GIS on the Internet, these are almost always from before 2012, and the URLs provided are now invalid. This initiative, commendable in its intentions, was also rather well thought out in its approach, particularly emphasizing dialog between stakeholders. Nevertheless, it was the first indication of the difficulty of applying GIS concepts and methods, which were technically and "culturally" designed to meet operational needs within clearly defined boundaries and domains, to the *territory* (in its most generic sense) or to *territories* considered in their heterogeneity.

5.2.2. *The mounting pressure from large communities*

It was with this in mind and at around the same time (around the 2000s) that large local authorities began to equip themselves with GIS, aided not only by the increasing accessibility and performance of computers and software, but also by the changes taking place in the territorial organization of France with the development of inter-municipal cooperation and the rise of the Regions (Feyt 2007). Keen to quickly and efficiently feed their GIS with reference data ("topographical data", land registry), these new actors increasingly and openly challenged the two dominant and indispensable institutions in terms of "information investment", namely, the IGN and the Directorate General of Taxes (now Direction Générale des Finances Publiques (DGFiP)), accusing them of hindering their projects by making access to crucial layers of information difficult and/or excessively costly. In his seminal 1999 report on *The prospects for the development of geographic information and the consequences for the IGN (Les Perspectives d'évolution de l'information géographique et les conséquences pour l'IGN)*, Congressman Lengagne did not hesitate to write that "the IGN had failed to recognize its own interests by denying potential users access to basic data" (Lengagne 1999).

This report also clearly set out findings and guidelines that would prove decisive for the development of what is now known as *sovereign geographic data* (Faure-Muntian 2018). Fundamentally, it already affirmed the strategic importance of geographic information for the state itself, indicating that "informational dependence would be the first step toward economic and political dependence". The report also noted that, "far more than simply changing the way maps are produced, the digitization of geographic information has completely transformed its uses". Indeed, within the large communities that were increasingly engaging in the implementation of GIS, practitioners now wanted to have information that was precisely tailored to their needs, no longer satisfied with the few, often obsolete and multi-thematic maps that had been available until then. At the same time, these same practitioners tended to a priori reject the use of a product that was too rich and therefore too complex and too expensive. The Lengagne report also emphasized that these increasingly diverse users wanted not only to enter their own data or that of their local partners, but also to take into account geographical locators other than those used by national reference systems. However, it noted that, despite this expectation and the integration of increasingly accurate and specific data, maps remained the preferred medium for presenting results and accessing information, albeit with one major change: these new tools now made it possible to produce a multitude of maps that met specific expectations and visions for the territory in question. This knockout victory of analytical and "automatic" maps over summary maps would resonate and at the same time contribute to the rise in number and power of increasingly specialized technical professions within large local authorities. Gradually, functional departments (sanitation, roads, environment, etc.) sought to equip themselves with their own specialized GIS, forcing GIS departments dedicated to a cross-functional approach based on geographic data to constantly rethink their positioning and added value.

5.2.3. *The revealed complexity of geographic data*

In defense of the IGN and the DGI, the projects to digitize their respective cartographic heritage proved to be as complex in concept as they were difficult to implement. Undertaken at around the same time, in the 1980s, the digitization of very large public alphanumeric databases (such as that of, for example, health insurance) was carried out relatively quickly and painlessly. Conversely, transcribing the content of a map into a data model combining geometry, topology, semantics and possibly semiology proved to be extremely complex, a challenge that was clearly underestimated at the outset. The development of the EDIGEO (Electronic Data Interchange standard in the field of GEOgraphic information) geographical data standard, which was supposed to become the "universal" exchange format in the sense that it would preserve as many properties of cartographic objects as possible, is illustrative. Its implementation and use proved so cumbersome that today the DGFiP is virtually the only organization using it to deliver cadastral data. Digitization costs may also have been a hindrance: for example, the "BD Topo" large-scale geographical referential in object format, available for a few departments in the mid-1990s, was provided at a rate of 2,000 francs/km^2, which for a department such as Hérault would correspond to approximately €3 million today.

Major regional geographic information projects: land parcels

The case of the land registry eloquently illustrates how the formal constraints of map digitization reveal different professional perspectives on the same object and lead to the production of different models. Based on the same historical paper cadastral documents, two cadastral databases from separate government agencies were developed, coexisted, and continue to do so. One, "ontologically fiscal" and "fundamentally accurate" (PCI), ensures that the actual (non-projected) surface area of parcels is represented. It is the product of the digitization of disjointed cadastral sheets covering irregular perimeters with sometimes geometrically random contours, but whose precise assembly was not historically a necessity. The other (land register database), which is strictly geographical, aims to establish the best possible correspondence between parcels and actual land use, covering the national territory in a homogeneous and continuous manner. It was not until 2014 that the *Single Cadastral Parcel Representation* (RPCU) project was launched as part of a partnership between IGN and DGFiP, which would provide a single description of the cadastre, but with an as yet uncertain timeframe (five departments completed in 2022).

The address

Launched later (in 2013), the project to create the *National Address Database* (Base Adresses Nationale (BAN)) reflects this same conceptual difficulty in defining a single, common geographical reference system. It aims to assign geographical coordinates to some 26 million addresses across the country that make sense for various uses and operators (government services, local authorities, Postal Service, etc.), some of which already had their own databases referring to different coordinates for the same address (middle of a plot, front door, electric meter, etc.). To date, 60% of addresses have been entered into the database through a partnership involving local authorities but also a participatory mapping project such as Open Street Map[1], reflecting a clear "cultural" and institutional shift compared to the 1990s and 2000s.

Box 5.1. *National projects for the Land Registry database and the Addresses database*

Furthermore, the digitization of geographic information has sometimes revealed or accentuated "cartographic oddities" that are not necessarily errors but simply the product of representations of "real space" created for a specific purpose, at a specific time, with specific means, constraints and methods. The historical administrative *mille-feuille* is also a cartographic *mille-feuille*. Each map or sheet taken in isolation, depending on what it was designed for, was coherent and satisfactory in itself; however, once digitized and georeferenced, its juxtaposition with its neighbor or its superimposition with a map based on a different logic can reveal the biases,

1 See: https://www.openstreetmap.fr/.

interpretations or approximations that presided over their development. This objectifying function of GIS (which is clearly its purpose and strength) has thus been a handicap, at least initially, by hardening the position of stakeholders who were already dealing with the internal task of digitizing their own data. In Box 5.1 and at the end of section 5.2, we will discuss three major examples – among others – of projects to develop national and, we might say, sovereign geographic reference systems, but which have found themselves, in one way or another, forced to contend with both the subjectivity inherent in cartographic representation and the diversity of uses and users.

These two examples reveal specific characteristics of digital geographic information that have gradually become apparent:

– Its most often "professionally situated" nature: the same object in the real world is viewed and described differently depending on the profession seeking to describe it.

– This same object is nevertheless most often considered, used, or processed by several different professional or institutional actors.

– These same actors frequently carry out their activities within an organizational, legal, cognitive, etc., framework that predates the digital age and with which they are required to work.

– The structuring and production of geographic databases are the result of both large-scale, lengthy and complex projects (conceptually, technically, financially, legally and even politically), but also collective and sometimes informal approaches to data and/or tools.

5.2.4. *Needs that are becoming more apparent, obstacles that are being removed...*

During the 2000s, geographic reference information, initially conceived on a national scale at the early initiative of the government, quickly began to reflect the actual or potential needs of local areas. This development can be explained by relatively independent factors:

– Starting with the 1999 Chevènement Act on *strengthening and simplifying intermunicipal cooperation*, major changes took place in territorial organization, notably with an increase in the number and powers and skills of intermunicipal bodies, mainly urban ones, which gradually discovered the value, and even the necessity of acquiring data and tools to understand their new territory and carry out new or enhanced tasks of an increasingly operational or strategic nature.

– Smaller local authorities are also getting involved in implementing GIS thanks to the increasing accessibility of software and hardware in all respects, and as national reference data becomes reasonably accessible.

– Regions, whose operational powers at local scale are fairly limited, need to build visibility and legitimacy, particularly by taking on a role in facilitating and coordinating local initiatives, where regional information can clearly be an asset.

– Technical agencies, particularly regional ones, dedicated to geographic information are gradually being put in place and are coming together, with a first seminar on *regional dynamics in geographic information* held in Poitiers in 2015.

– These *new territories* are organizing themselves to exert political and technical influence at the national level (with, for example, the AdFC[2] or the Association des Régions de France (ARF)[3]).

5.2.5. *...and accelerators being placed*

The importance of what was not yet called *Open Data* had been highlighted in France as early as 1999 in the Mandelkern report on *the dissemination of public data and the digital revolution* (Mandelkern and Du Marais 1999). The report already devoted a specific section to geographic information. However, this section was part of a chapter that also dealt with legal data and business data, entitled "Three sectors have emerged as particularly contentious" and beginning with the sentence "The National Geographic Institute is highly contested".

At the European Community level, the movement launched around the reuse of public data led in 2003 to an initial directive, which was translated into law in France in 2005. In the specific field of geographic information, the European working group GINIE (Geographic Information Network in Europe) produced a report in 2003 recommending the establishment of Spatial Data Infrastructures (SDI) guided by highly pragmatic principles and objectives that prioritize operational uses and usability.

A key milestone was reached in 2007 with the European INSPIRE (INfrastructure for SPatial InfoRmation in the European community) directive

2 Assembly of French intermunicipalities, now known as Intercommunalités de France (https://www.intercommunalites.fr/).

3 Association of French Regions (https://regions-france.org/).

establishing a geographic information infrastructure in the European Community, which was transposed into French law in 2010. Public authorities (a concept that will remain rather vague) will thus be required to provide access to a set of data relating to environmental issues in the broadest sense. As will be the case a little later with Open Data, this obligation with its somewhat vague terms of application will greatly facilitate or legitimize the circulation of data that had already often informally taken place. In fact, as highlighted in the 2022 European evaluation report[4], although dedicated to the environmental field, the INSPIRE Directive has been invoked for the implementation of other territorial policies, even in the absence of legal requirements.

However, before going any further into these institutional dimensions, it is difficult to ignore what, in 2006, literally constituted a breakthrough in the rather small world of digital geographic information, which at the time was focused on the issue of access to sovereign reference data: the launch of Google Maps in France. Suddenly, what had previously been reserved for a professional community, namely, access to digital representations of the territory, even if only through aerial photography, became possible for anyone with Internet access; this would even quickly become not only commonplace but indispensable. Of course, this did not prevent geomatics from continuing to make steady progress, as it was a completely different matter; but, at least in the early days, GIS project managers sometimes had to work hard to convince administrative or political decision-makers who did not necessarily understand why they should pay and reorganize for something complex and expensive, which could be obtained easily and free of charge from home. Nevertheless, this "disruption" may have helped to develop an appetite among local practitioners for access to geovisualizations and then to geographic data itself via web mapping tools.

In fact, over the years, the implementation and management of these tools has become an increasingly important part of the daily work of geomatics specialists, as already indicated in the survey conducted jointly in 2013[5] by the Association française pour l'information géographique (AFIGEO) association[6] and the GeoRezo[7] and MAGIS (Méthodes et Applications pour la Géomatique et l'Information Spatiale)[8] networks, which we will return to later.

---

4 2022 evaluation of the INSPIRE Directive (https://cnig.gouv.fr/evaluation-2022-de-la-directive-inspire-a25974.html).

5 Survey on the evolution of geomatics professions, 2013 (https://afigeo.asso.fr/wp-content/uploads/2021/01/05/afigeo-colloque-formation-metiers-2013-session1-metiers.pdf).

6 French Association for Geographic Information (https://afigeo.asso.fr/).

7 See: https://georezo.net/.

8 See: https://gdr-magis.cnrs.fr.

5.2.6. *Geographic information as a "public good"*

These various processes will gradually lead to a form of "age of reason" in the relationship between Territory and territories around digital geographic information based on two complementary concepts: subsidiarity, with the recognized benefit of a clear division of roles, and the idea of the common good, with the imperative of information governance. Thus, following the example of the National Institute for Statistics (Institut National de la Statistique et des Etudes Economiques (INSEE)), which initiated an open data policy in 2003, the open French public data platform[9] was launched in 2011, with more than 40,000 datasets now available. In 2017, the Directorate General of Public Finances (DGFiP) and the Etalab[10] mission provided access to data from the computerized cadastral map[11]. In 2021, the IGN opened up access to its public data, the culmination of a process that had lasted nearly twenty years and whose extension initiated an approach based on the concept of *geocommons*[12].

These "technical" advances reflect a broader shift in the relationship between the state and its territories, both in terms of its relationship with large urban communities that are increasingly capable of "informational empowerment", and in terms of the functional and symbolic imperative to continue to "represent the territory" through cartography and, now, geographical reference data. These intentions are clearly set out in the report on *Sovereign Geographic Data* (Faure-Muntian 2018) published in 2018 under the responsibility of Congresswoman Valéria Faure-Muntian. While highlighting the pivotal role of the IGN, the Prime Minister's mission statement calls for the scope of sovereign geographic data – in other words data necessary for public action – to be defined, but also for methods of collaboration and, more broadly, governance between state services and local authorities to be proposed. The report immediately highlights the risk of "disruption" (no doubt to avoid talking about falling behind) of public authorities in relation to GAFAM[13] in the provision of geolocated information, explicitly referring to the warnings made in the Lengagne report twenty years earlier: "the persistence of these questions betrays the difficulty that public authorities have had in adapting their organization and methods of intervention in a rapidly changing context, marked by numerous and scattered initiatives, as well as rapid and continuous technological progress".

---

9 See: https://data.gouv.fr/fr/.

10 See: https://etalab.gouv.fr/.

11 See: https://cadastre.gouv.fr/scpc/accueil.do.

12 See: https://ign.fr/institut/la-demarche-geocommuns.

13 Acronym for the "web giants" (Google, Apple, Facebook, Amazon, Microsoft).

Nevertheless, perceptions and perspectives have clearly changed: noting "the extreme variety of geographic data involved in public policy decisions and the complexity of the ecosystem of stakeholders", the 2018 report outright rejects the idea of a monolithic approach, as well as, on a technical level, the development of standards and, in terms of governance, the establishment of a single model. Faced with the geolocation services offered by GAFAM, the challenge is to prioritize "assumed" and "enforceable" data (land registry, urban planning documents, etc.) as well as generic layers describing the national territory, notably through the Large-Scale Reference System (Référentiel à Grande Echelle (RGE)). Conversely, the report notes that business data are increasingly being produced by (or for) public or semi-public bodies responsible for multiple territorial policies (security, environment, land use planning, transport, health, agriculture, etc.). The central issue is therefore how to coordinate the production and circulation of all these data. That said, the report incidentally mentions that some of the individuals interviewed described the production of data by the various public services as a "Brownian motion", suggesting that it would be futile to try to organize it.

5.2.7. *Conditions in place for widespread adoption of local GIS?*

In the 2000s, during the maturation phase of GIS as a tool and organizational device, local initiatives may have been hampered by institutional, material or financial constraints, but this would no longer be the case from the 2010s onward. The capacity to produce specific data, particularly through geolocation, is becoming increasingly significant, diverse, accessible and interoperable. Ever more powerful open-source software can be installed on increasingly powerful and inexpensive computers. Digital technology, including geovisualization, is at the heart of many professional and individual practices, and new local practitioners have, in most cases, received training in geomatics.

From the perspective of the sphere of digital geographic information, the job has been done, and done rather well, even if it took a little time. All that remained was to roll out or adapt the approach and methodologies designed primarily for large urban communities across all territories, naturally in a differentiated manner according to size, resources and geographical context. However, this did not take into account France's unique territorial organization. Furthermore, it was known that French territorial structuration would have to face profound and diverse changes during the 2010s, in which issues relating to geographic information, and even territorial knowledge, would generally not be a priority.

5.3. Territories that are structurally unequal in their capacity for self-knowledge

As indicated in the Introduction, we will not attempt to define or discuss the concept of territory, which is a scientific issue in itself. The perspective here is purely functional: the aim is to try to understand how GIS tools for managing and understanding territory have been able – or not – to become established within the historical or new structures responsible for "basic" territories, namely, local authorities, municipalities and inter-municipal bodies. The territory is thus conceived as a formalized framework for professional practice and political action. However, to grasp its heterogeneity, and even its complexity, it is necessary to review this territorial organization at the national level, understood from the perspective of the needs and capacities of these various authorities to produce specific knowledge about their own territory.

5.3.1. *Disparities in territorial structure*

In 2022, France had almost 35,000 municipalities, representing 40% of all municipalities in the European Union. Just over half have fewer than 500 inhabitants and 98% have fewer than 15,000. Since the end of the 19th century, this peculiarity, which is clearly also a functional challenge, has been addressed by seeking to encourage and ultimately impose the grouping of municipalities through inter-municipal cooperation. Since the 1990s, a series of increasingly prescriptive laws have been introduced. The most recent, the NOTRe Law of 2015 on the *New Territorial Organization of the Republic*, made it mandatory for all municipalities to join an intermunicipal structure, but established several categories of *Public Intermunicipal Cooperation Institutions* (Etablissement Public de Coopération Intercommunale (EPCI)) defined mainly according to urban structure and number of inhabitants. Each of these types of EPCI has been assigned mandatory or optional powers, which are, in principle, complementary to those of the other territorial levels operating in the same territory (municipalities, departments, regions).

It is at this territorial level of EPCIs that we will focus in particular when discussing the penetration of GIS in the territories, for several reasons:

– First, because the 18,150 municipalities with fewer than 500 inhabitants in 2019 share 29,000 local government officials or 1.6 officials per municipality; and even though there are 36 employees per municipality for those with between 2,000 and 5,000 inhabitants, having a GIS is not even an issue in view of the resources available and the daily and vital tasks to be performed.

– Second, because the typology of EPCIs makes it possible to take into account the disparity between municipalities, with INSEE able to identify and characterize families of EPCIs by distinguishing, for example, between *rural, peri-urban and autonomous communities*.

– Finally, in terms of purpose, because in 2020, the Local Authorities Code entrusted inter-municipal bodies with strategic functions, including for communities of municipalities, the obligation to develop "a joint spatial development and planning project"; for urban communities and metropolitan communities, "a joint urban development and land use plan"; and for metropolitan areas, "a plan for economic, ecological, educational, cultural, and social development and land use". Underlying this last point is the clear interest, if not necessity, of having cross-functional and integrated information capabilities (Feyt 2010). In fact, we will see later that it is mainly public establishments for intermunicipal cooperation (EPCI) that seem to be active in this area.

After many changes, the most significant and recent of which was a very large reduction in the number of EPCIs (2,601 in 2009, 1,258 in 2019), which automatically led to an increase in their population and surface area. We can assume that the situation has stabilized overall, at least for a number of years. To date, there are four types of EPCI in France, with very different characteristics, which are presented here in schematic form:

– *metropolitan areas* (22 including Greater Lyon) and *great urban communities* ("C.U.", 14) bringing together municipalities in large urban areas;

– *urban communities* ("C.A." (Communauté d'agglomération), 207 in mainland France), which, except in special cases, must have at least 50,000 inhabitants and a city with more than 15,000 inhabitants;

– *communities of municipalities* ("C.C.", 995) covering the rest of the territory, with minimum population thresholds that are constantly being lowered to take account of very sparsely populated areas, forcing them to group together a considerable number of municipalities over extremely large areas.

Other "trans-EPCI" structures exist, such as *metropolitan regions* and *regional nature parks*, but only the types of EPCI mentioned above cover the entire national territory.

In terms of areas of action, the metropolitan areas, the urban communities (C.A.) and the most populous community of municipalities (C.C.) (one-third of which have more than 100,000 inhabitants) exercise powers in the domain of management (waste, sanitation, etc.), planning (roads, public transportation, etc.), urban policy (urban planning, housing, etc.), economic development, environment and quality of life. All of these tasks require significant human resources and diverse expertise. Non-urban agglomeration communities and communities of municipalities have fewer powers under the regulations, but are often called upon to take on functions such as sanitation, for example, in place of municipalities that are increasingly unwilling and/or unable to do so themselves. However, these areas, which can be described as peri-urban, peri-rural or rural, face problems and challenges in areas that are just as diverse and complex as those faced by urban agglomerations. These issues also require the ability to diagnose, simulate, monitor, evaluate, inform, etc. In short, they need diverse territorial data and the ability to maintain and analyze it.

5.3.2. *The implications in terms of capacity for action*

The importance and nature of the resources available to local authorities are largely linked to the size and structure of their population; thus, the grants allocated by the state are essentially calculated on the basis of demographic and fiscal criteria. The concept of surface area is virtually never taken into account; however, when geographical databases are created and maintained, their spatial extent is not an insignificant constraint. The land registry is just as fragmented in rural areas as it is in cities, and the issues and disputes that may arise from it are just as problematic. This is even more true when it comes to physical or environmental data, and while the density of networks is lower than in urban areas, it is just as important to know their extent and condition in rural areas.

Thus, as Table 5.1 shows, if we consider only demographic factors, we can see that there is relative equity between the three main types of EPCI: each accounts for around one-third of the French population, and when we compare the number of local government employees to this population, based on figures from the national center of local public officials (Centre National de la Fonction Publique Territoriale (CNFPT)[14]), the ratio is certainly in favor of large urban areas, but within reasonable proportions. Conversely, if we consider the types of EPCI in terms of surface area, the difference appears considerable: large urban intermunicipal communities cover 3% of the mainland France, while communities of municipalities cover 78%. Admittedly, the ratio of "number of employees per square kilometer" is meaningless in itself, but the enormous difference illustrates, almost absurdly, the failure to take spatial extent into account when allocating resources.

---

14 National summary of reports to the technical committee on the state of local authorities, CNFPT report, 2019 (https://www.cnfpt.fr/s-informer/nos-actualites/le-fil-dactu/synthese-bilans-sociaux-collectivites/national).

	Total	Metropolitan area + large (UC)	Urban communities (CA)	Community of municipalities (CC)
Number of inhabitants (in millions)	**6.54** **100%**	20.8 32%	22.1 34%	22.5 34%
Area (km^2)	**552,000** **100%**	14,000 3%	105,000 19%	433,000 78%
Number of agents	**254,000** **100%**	85,000 33%	87,000 34%	82,000 32%
Number of agents per 1,000 inhabitants		**4.1**	**3.9**	**3.6**
Number of agents per square kilometer		**5.9**	**0.8**	**0.2**

To avoid taking into account extremely large territories, particularly French Guiana, the analysis focused solely on metropolitan public establishments for intermunicipal cooperation (EPCI), estimating and deducting the number of civil servants in the overseas departments based on data from the CNFPT.

Table 5.1. *Metropolitan France 2017. National statistics for different types of EPCI (processed using INSEE and CNFPT data)*

Communities of municipalities and some urban communities therefore have a lot of space, and often even a diversity of spaces, to take into account when developing and implementing their territorial policies. At the same time, they have fewer human resources, both in absolute and relative terms, while having to carry out increasingly diverse and pressing tasks. In this context, even when aware of the benefits of doing so, which is not always the case, deciding to implement a GIS can prove problematic, especially for EPCIs in charge of very large natural areas requiring the consideration of complex data (floods, fires, biodiversity, etc.).

5.4. The dissemination and place of GIS within territories

As indicated in the Introduction, the subject that interests us here is not "geomatics for territories". The entire book would not be enough to cover all of the possible uses of digital geographic data and geomatics for addressing the issues faced by territories, from network management to archaeology, mobility and pastoralism. On the contrary, while the potential is endless and diverse, we must first examine the very conditions for the existence of a GIS on and for the territory. This GIS must include layers that are both specific and diverse; in other words, it must enable self-knowledge and provide added value compared to processing based solely on national reference data.

5.4.1. *Sources of information*

In practice, it is not easy to obtain a precise and comprehensive understanding of the dissemination and distribution of GIS within French local authorities in recent years, as no census or survey on the subject is available. However, there was a marked interest in monitoring the dissemination of GIS quantitatively and qualitatively in the 2000s and early 2010s. Data are available from the GIS Observatory set up between the late 1990s and early 2000s by the GIS consulting firm IETI and mentioned in Boris Mericskay's master's thesis (Mericskay 2007). At the same time, government departments had set up a similar system for internal use. For its part, the AFIGEO association conducted a national survey of practitioners identifying as geomatics engineers in 2003, 2005 and 2013, even though, as we shall see, this could cover a wide range of professional practices.

In addition to these elements, which enable us to assess developments and continuities, two "collateral" sources of information were used in an attempt to evaluate the current situation in the absence of specific surveys:

– The first, from a qualitative and comparative perspective, relates to the nature of the human resources mobilized within local authorities around issues relating to geographic information.

– The second, quantitative and historical, focuses on changes in the provision of consulting services and equipment to local authorities, as reflected in calls for tenders in the context of public procurement.

The first source of information identified consists of analyzing basic information (job title, type of community) available in the member directory of the CNFPT's Geographic Information e-community, a national organization responsible for training 1.9 million local government employees and monitoring changes in the many professions involved. Created in September 2023, this forum already has almost a thousand members. Although it does not necessarily include all local government employees interested in geographic information, this directory can be considered a relatively reliable source of information on the types of positions and local authorities concerned. The second source of information is the data provided by the Official Bulletin of Public Procurement Notices (Bulletin Officiel des Annonces des Marchés Publics (BOAMP))[15]. Using the keywords "geomatics" and

15 See: www.boamp.fr.

"geographic information system", the site extracted 620 notices or award notices between 2006 and 2023, involving amounts of €15,000 or more, depending on the year. It is therefore possible that small contracts may have been omitted from the corpus, but in view of the results, it can be considered indicative of the overall trend and structure.

In a necessarily simplified manner, these different sources of information highlight three distinct and, it seems, persistent characteristics.

5.4.2. *The predominance of public affairs...*

With 1,820 respondents across France, the 2013 AFIGEO survey clearly shows the predominance of public sector work, with 65% of respondents working in the public or semi-public sector. It can be assumed that a significant proportion of the work carried out by the 27% working in private practice or as private sector employees is related to public sector contracts. Moreover, this share of the public sector grew by nearly 20% between 2003 and 2013, reflecting the recognition of the interest, or even the necessity, of internalizing GIS as a permanent, if not strategic, function. This predominance of links to public action is confirmed by the initial results of the *economic study of the geonumerical ecosystem* published in 2023 by the AFIGEO[16] association, which showed that 85% of the 57 companies that responded to the online survey said they worked for the public sector.

5.4.3. *...and of geo-the-maticians*

With regard to initial training, the 2013 AFIGEO survey on the geomatics profession reveals that more than three-quarters of respondents are university graduates. In terms of fields of study, thematic approaches (geography, land use planning, life and earth sciences) dominate with more than 60%, followed by specific training in geomatics (few at the time) and computer science, accounting for 20%. On the contrary, training in what could be called "large scale" (surveying, topography, photo interpretation) and even cartography accounted for less than 10% of respondents. Whether these characteristics were included in the initial job description or have gradually become more prominent, this disciplinary eclecticism clearly reflects the diversity of functions and positions. An analysis of the current training offer suggests that this predominance of thematic approaches over strictly technical approaches to geomatics not only persists but has been reinforced; in 2024,

16 Economic study of the geonumeric ecosystem, AFIGEO, 2023 (https://www.afigeo.asso.fr/wp-content/uploads/2023/07/03/afigeo-rapport-enquete-presentation-du-20-juin.pptx.pdf).

of the 24 master's degrees listed on the government's master's degree information website[17] under the keyword "geomatics", seven fell exclusively within the Science, Technology and Health sector, with all of the others (including three "hybrid" degrees) falling within the field of Humanities and Social Sciences.

Similarly, information on members of the CNFPT's Geographic Information e-community shows that more than two-thirds do not have a job title directly related to digital geographic information, with this figure rising to 90% for municipalities, as shown in Table 5.2. The EPCIs, which have the largest number of members, are an exception, which tends to reinforce the idea that it is primarily inter-municipal communities that are adopting GIS.

	Yes	No	Workforce	%
EPCI	49%	51%	**340**	**38**
Municipality	10%	90%	**284**	**32**
Public institution	25%	75%	**101**	**11**
Department	35%	65%	**96**	**11**
Union	38%	63%	**48**	**5**
Region	32%	68%	**28**	**3**
Total	**31%**	**69%**	**897**	**100**

Table 5.2. *Job titles of members of the CNFPT geographic information e-community related (or not) to geographic information (GIS, geomatics, cartography)*

This high percentage of "non-specialists" echoes the 2013 AFIGEO survey. Only a quarter of respondents said they devoted all of their working time to geomatics, around one-third devoted between 75% and 100%, with the remainder distributed fairly evenly between 75% and 0% of their activity.

Returning to the current situation and looking at the professional profiles of the "non-geomatics" members of the CNFPT e-community, the "(very) diverse" category is by far the most dominant. It covers the entire range of local government professions, from municipal police to management control, road maintenance and early childhood education. As Table 5.3 shows, two areas stand out: urban planning and land law, closely linked to the land registry, and digital technology and IT services, with which GIS has a necessary relationship and sometimes even a hierarchical link.

17 See: www.monmaster.gouv.fr.

	(Very) diverse	Urban planning/ development	Land law	Digital	Environment/ risks	Workforce
EPCI	69%	20%	7%	3%	2%	**256**
Municipality	58%	16%	12%	8%	6%	**172**
Public institution	82%	9%	3%	4%	3%	**76**
Department	71%	8%	0%	19%	2%	**62**
Union	77%	10%	3%	7%	3%	**30**
Region	68%	5%	5%	11%	11%	**19**
Total	**68%**	**15%**	**7%**	**7%**	**3%**	**100%**

Table 5.3. *Areas of activity of e-community members who do not have a role directly related to geographic information*

5.4.4. *Territorial disparities*

As early as 2005, the *IETI GIS observatory* revealed that 90% of municipalities with more than 100,000 inhabitants and 100% of urban communities had begun implementing a GIS. This was at a time when inter-municipal cooperation was only just beginning to take off and had to resolve political, organizational and financial issues before turning its attention to knowledge tools. Nevertheless, between 2000 and 2005, metropolitan communities (with 50,000 inhabitants or more) saw their GIS equipment rate rise from 13% to 44% (based on the current number of metropolitan communities), while urban communities reached 100%. Municipalities with fewer than 5,000 inhabitants, which are by far the most numerous, had an equipment rate of 4% in 2005, and it is not clear how this has changed since then.

A few years later, the 2013 AFIGEO survey revealed that just over 60% of responding geomatics specialists worked in organizations with more than 100 employees. Bearing in mind that the AFIGEO directory of geospatial companies indicated that in 2015, 92% of private companies in the sector had fewer than 100 employees, it is logical to assume that most of these employers are public entities. Furthermore, a comparison with CNFPT data on staff numbers by type and size of ECPI shows that with around 100 or more employees, the average number of employees per community of municipalities (81 in 2019) is exceeded. Employment structures are therefore mainly urban and metropolitan communities, or municipalities with 10,000 or more inhabitants, representing 3% of French municipalities.

Based on Table 5.4, an analysis of some 240 job titles explicitly referring to the field of GIS or geomatics among members of the CNFPT's Geographic Information *e-community* allows us to formulate a number of hypotheses regarding the structure of activity within the local authorities concerned.

	Director	Head of department, head of unit or division	Head of mission	Project manager, GIS manager	Operational functions	Total
EPCI	4%	23%	6%	32%	35%	**100% (168)**
Municipality		7%	7%	54%	32%	**100% (28)**
Department		21%	12%	38%	29%	**100% (34)**
Region		11%	11%	33%	44%	**100% (9)**
Total	7	**49**	**17**	**85**	**81**	**239**

Table 5.4. *Categorization of the job titles of members of the CNFPT geographic information e-community working explicitly in the field of GIS and geomatics*

The most striking feature is clearly that, across all functions, EPCIs are overrepresented compared to other local authorities. It is also interesting to note that of the seven members with management roles (all in urban local authorities with more than 150,000 inhabitants), five have job titles that refer to geographic information in a generic way: *[data and]/[resources in] [geographic information],* suggesting that geomatics and GIS have become tools within a broader range and perspective. Furthermore, it is impossible not to notice, particularly for EPCIs and municipalities, the very high proportion of "GIS managers" or "GIS project managers", especially when compared to operational functions (administrator, geomatics specialist, cartographer, etc.). It should be noted that, in the classification, care has been taken to distinguish these functions from those of "GIS department/ unit/cell managers". It is thus reasonable to assume that these managers are primarily responsible for themselves and that the GIS function often relies on a single person. It is also interesting to note the relative importance of the role of "GIS project manager", which reflects the significance of organizational issues.

These different elements paint a picture of "territorial geomatics" in which two typical configurations coexist. On the one hand, there is a GIS department or division operating as a resource within a large urban community, run (among other professions) by dedicated geomatics specialists. On the other hand, there is a medium-sized city or urban community with a dedicated GIS officer for all or part of the GIS. While the existence of a gradient between these two "extreme" configurations may be intellectually tempting, the hypothesis is rather that there are threshold effects and that, once operational or specific data integration is undertaken, the needs and challenges become such that they require the long-term acquisition of additional human and material resources…and therefore the capacity to do so. In the opposite case, it is still possible to integrate reference data that are now easily accessible but essentially standardized and therefore, as already mentioned, with little added value in operational terms or in terms of contributing to self-knowledge. In this scenario, the function has to contend with a difficult or even precarious position within the structure because it is potentially optional or can be outsourced.

5.4.5. *A "threshold effect" in the territorial diffusion of GIS?*

It seems important to test both thc typology proposed above and this hypothesis of a threshold effect, below which it seems difficult to implement or even apply the GIS principles and approaches developed in, by, and for large urban communities.

In this case, analysis of public tenders issued from the BOAMP provides some insight. Figure 5.1 shows the change in the number of tenders published between 2006 and 2023, distinguishing between types of public structures (from national to local), with (inter)municipal structures represented with the lightest grays. Two clear patterns emerge: the chronological trend, with a peak in the period 2006–2013, followed by a rapid and steady decline affecting all categories, reaching a very low level in 2023, and the weight of EPCIs (public establishments for intermunicipal cooperation) compared to cities, with a particular predominance of large urban areas.

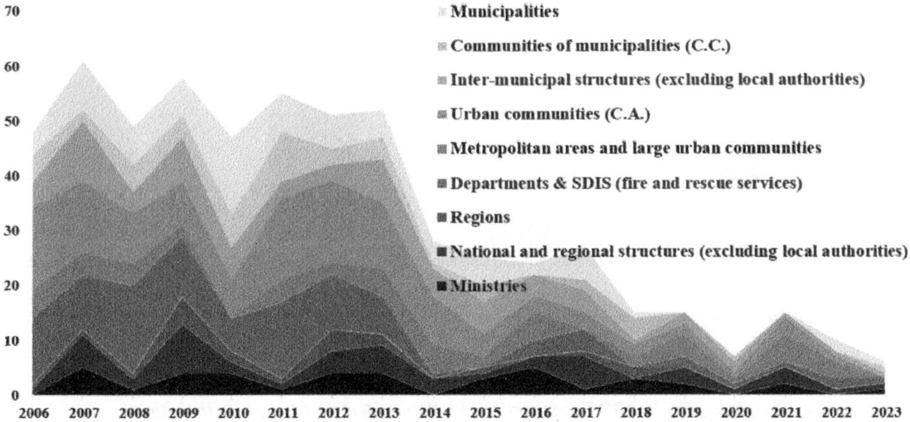

Figure 5.1. *Trends in calls for public tenders relating to GIS and geomatics from the Official Bulletin of Public Procurement Notices (BOAMP)*

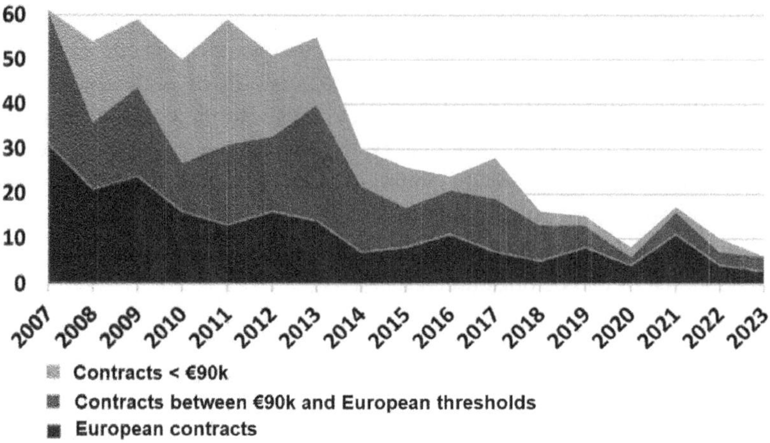

Figure 5.2. *Trends and distribution of public tender prices (BOAMP)*

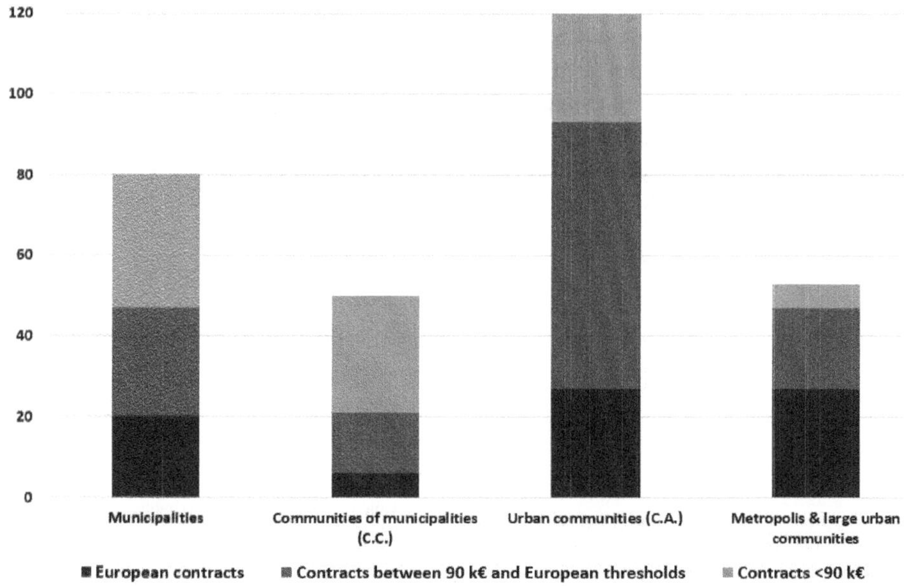

Figure 5.3. *Breakdown of types of public procurement over the period 2007–2023 according to the nature of local authorities*

Figures 5.2 and 5.3 illustrate this trend in terms of market size and differences between types of local authorities in the (inter)municipal bloc. Here again, communities of municipalities are the least represented in terms of both the number and the value of contracts.

The question that then arises is how to determine the proportions of local authorities appearing in the BOAMP database in relation to their respective categories. As previously mentioned, the overall number of EPCIs was significantly reduced during the 2010s, so we chose to take into account the "arrival" figures for 2023 (Figure 5.4).

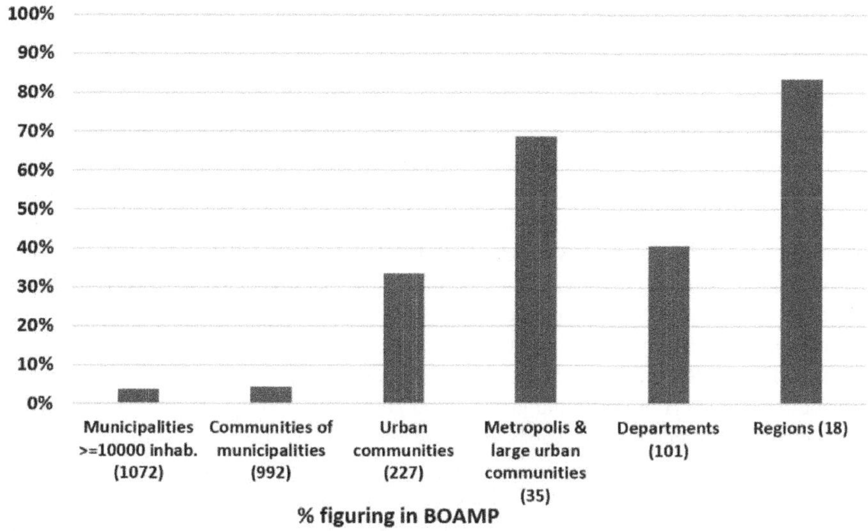

Figure 5.4. *Share of local authorities (2023 workforce) appearing in calls for tenders between 2006 and 2023 (BOAMP)*

Publication in the BOAMP is not mandatory for contracts below the European threshold, and the notice may be published in other legal gazettes; these figures should therefore be viewed primarily from a comparative perspective. However, the differences are significant enough to be considered meaningful. Only a few percent of the thousand municipalities with more than 10,000 inhabitants and the thousand communities of municipalities have commissioned digital geographic information services over the past fifteen years. This represents about one-third of urban communities with 50,000 inhabitants or more. It is certainly possible today to acquire software and reference data at low cost, as well as to recruit an environmental or urban planning project manager with expertise in geomatics who can also serve as a "GIS manager". However, even on a "scale model", this makes it difficult to build, maintain and develop geospatial resources based on the model used by territories above the “threshold”. These rates, even if underestimated, combined with the significant decline in the number of calls for tenders, seem not only to validate the hypothesis of a threshold in terms of local authority size, but also to show that the phase of structuring GIS services by local authorities capable of doing so is nearing completion.

Major regional geographic information projects: the PCRS

The national project to implement the Simplified Street Map Plan (Plan de Corps de Rue Simplifié (PCRS)) clearly illustrates the various levels of complexity involved in digital geographic information. This is particularly obvious when it comes to producing a territorial reference system that must comply with regulatory and partnership requirements and meet a wide range of public needs. Basically aimed at providing a base map enabling public and private actors to have an accurate representation (less than or equal to 10 cm) of the location of all networks to avoid damage to structures, the principle of implementing the PCRS was established under a law on damage to networks that came into force in 2012. Technically, the PCRS can exist in two forms, with increasing performance and cost: images via very high-resolution aerial photography, and vectors integrating the modeling and representation of objects that make up the road network and networks. Designed to be implemented on a partnership basis, the PCRS was the subject of a national agreement ratified in 2015 by state (IGN), regional and socio-professional (AMF, ARF, ADCF, etc.) and major energy network operators (Électricité Réseau Distribution France (ENEDIS) and Gaz Réseau Distribution France (GRDF)). The PCRS creation projects in and by the regions also give rise to local protocols.

Ten years after the formal launch of the initiative, the results have been mixed, despite the efforts of a national coordinating body with limited human resources and the involvement of the National Agency for Territorial Cohesion (ANCT, Agence Nationale de Cohésion des Territoires). To date, according to data provided by the ANCT, 23 metropolitan areas or urban communities already have a PCRS or are in the process of developing one, with nearly half of them already engaged in an update phase. Approximately 40% of the projects involve vector or mixed-mode acquisition. Of the 223 French urban communities, around twenty are in the initial implementation phase and five are undergoing updates, with around 40% also using vector or mixed mode. As far as municipalities are concerned, half a dozen large cities – including Paris – and an even smaller number of communities of municipalities have launched initiatives. For the rest of the country, it is largely the departments that are taking the lead and sometimes with difficulty. Given the size of the areas involved, they usually proceed in stages, and it is not easy to know how much land has actually been processed. In any case, to date, 63 departments (out of 101) have undertaken to carry out a PCRS on all or part of their territory. Among them, nearly 10% are already engaged in updating projects, and slightly more than half have completed or are in the process of completing the initial creation phase(s), with three-quarters of the projects focusing on the image mode.

Therefore, the 2026 deadline initially set for urban areas seems unlikely to be met. There are several reasons for these difficulties:

– First of all, the financial aspect: the cost of creating the PCRS vector for all built-up areas in a department can easily reach €5 million. The image version, which is the only reasonably feasible option for rural areas, is significantly less expensive but much less effective in operational terms. Added to this is the obligation to update the system.

– The partnerships aspect, insofar as the financial arrangements can be complex and difficult to negotiate, particularly with network operators about the distribution keys or even on the very principle of their participation, arguing that public data are free of charge due to its openness.

– The technical aspect, where data already exist at operators but are based on models that do not comply with or are incompatible with the PCRS standard.

Box 5.2. *The national project to implement the Simplified Street Layout Plan*

The PCRS's current work once again highlights the differences not only in practice but also in potential between dense urban areas and rural areas. It could certainly be argued that the density of networks and uses (or users) that depend on them is less significant in rural areas and small towns than in urban areas. However, on their own scale at least, the consequences of potential damage can be just as penalizing. This vulnerability, linked to a lesser capacity for self-knowledge, constitutes, at least comparatively, an additional structural handicap.

In this context, a real problem or moral dilemma arises for both the "geomatics world" and the "territorial world": How do we deal with this clearly growing imbalance between the geospatial resources of large urban areas and those of territories that are "small" in socio-demographic terms but "important" in geographical terms and equally confronted with complex and diverse challenges? However, as already pointed out, GIS are designed to deal with space as much as with spatial functionalities. In this area, as in others, there is a growing gap in terms of access to services between urban areas with increasingly significant material, human and cognitive resources, and areas of medium or low density that are forced to "do with or without". This observation is unfortunately commonplace and often highlighted – for example, in relation to access to healthcare or higher education. It is much less common when it comes to the ability of regions to acquire self-knowledge. However, just as with access to healthcare, we should not be satisfied with what can only be described as a decline. This is particularly true given that, just a few decades ago, the level of cartographic description was more or less identical across the entire country, at least in mainland France.

5.5. Territorial data: a new frontier for geomatics specialists?

While geomatics remains largely a subject for "insiders" within the local government sector, the same cannot be said for data. Even without GIS equipment, local authorities of all sizes have fully embraced the digital age. The challenge now is to figure out how to leverage the vast amounts of data they produce or exchange with their partners – primarily government agencies – but also how to address the challenges and obligations associated with open data for those with more than 3,500 residents. Indeed, while only 16% of the local authorities concerned currently comply with the obligation to open up data, beyond the disparity in size and resources, the problem seems to be much more widely linked to a lack of interest in data on the part of elected officials who are called upon and mobilized on an increasing number of diverse issues. Even in the digital sphere, the subject of data takes a back seat to issues such as digital regional development, user services, inclusion, cybersecurity, etc. These observations and questions prompted the government to set up the Data and Territories Mission, whose report was published in September 2023 (Hennion et al. 2023).

In this report, data are described as *geographic data, data produced as part of public statistics, or even real-time data from sensors*; in other words, almost entirely digital geographic information. Interestingly, without promoting or even mentioning geomatics as such, the report repeatedly refers to the benefits of relying on *geomatics specialists*. Open Data thus appear to be a new, legally sound and politically valuable way of thinking about the issue of territorial information and its value. In this context and from this perspective, geomatics specialists potentially have a key role to play. As the report acknowledges, within local authorities, "apart from a few very specific professions and roles related to geographic information systems – or not – there are few areas where data skills are a prerequisite". While noting with satisfaction the opening up of its databases by the IGN, the *Data and Territories* mission report finds that access to data remains difficult for both technical and "cultural" reasons. Among these are the significant data organization in silos produced by the state and its operators, with a need to improve its "discoverability"; the existence of "ambiguous duplicates"; and interoperability issues between national repositories and local databases. On this last point, once again implicitly highlighting geomatics, the report emphasizes that geolocation is clearly a key cross-referencing tool.

Finally, echoing the recommendations of the report on sovereign data, the *Data and Territories* report (Hennion et al. 2023) highlights the shortcomings or imbalances between central and decentralized government services and local authorities in terms of territorial information governance. Here again, geographic information serves as a “model”. The report states that “if there is one area where governance between local authorities and the State can serve as an inspiration, it is that of geographic information”. Admittedly, the reference is mainly to regional platforms. Nevertheless, in terms of understanding and resolving differences in cultures and professional practices surrounding the use of digital information, since the 2010s, geomatics specialists have not only contributed to experimentation and testing, but also to building original and recognized transferable skills and knowledge.

A fairly similar vision was developed in the 2018 report on *The digital transformation of territories: challenges, levers, and action plan for intermunicipal communities*[18], produced by the Association of Chief Executives of Local Authorities in France, based on a survey of more than 130 chief executives representing the diversity of intermunicipal communities. Admittedly, the overall issue at stake was the place and role given to digital technology, not only in the functioning of the community but also in that of the territory itself. In fact, when dealing with data relating to the territory, it can be considered that most of these data have a geographical dimension; however, this does not necessarily justify or proceed from a GIS. This is undoubtedly where geomaticians, like intermunicipal authorities, have a “part to play”. The *Association des Directeurs Généraux des Communautés de France* (ADGCF, Association of Chief Executives of French Communities) survey shows that 85% of EPCI directors consider that it is up to the intermunicipal authority to lead the project of digitizing the territory. They do so on behalf of municipalities that individually do not have the capacity to do so, and with a view to ensuring consistency in terms of the space in which residents and businesses operate. However, 90% of them consider that they do not have a clear roadmap because of budgetary constraints and/or a lack of motivation on the part of elected officials, and because their EPCI still has to “digest” the changes mentioned above. Here too, however, the GIS is clearly highlighted as a key contributor: among the five digital projects considered priorities by the ADGCF, the GIS ranks first. GIS’s ability and legitimacy to share its expertise and experience, particularly its cross-functional nature, with other areas of digital use are also highlighted in the 2023 edition of the *National barometer on data in the regions* by the *Data Publica*[19] observatory. The survey also highlights that the insufficient data culture and lack of skills are the main obstacles to the development of data use, noting that GIS services, due to their long history and expertise, often play a leading role in data management in general.

---

18 See: https://www.adgcf.fr/upload/billet/739-lb_digitalisation-des-territoires_v4-o.pdf.

19 See: https://observatoire.data-publica.eu/.

These "foot calls" to geographic information practitioners are clearly not intended to make GIS tools "machines that know and do everything, everywhere". In fact, the appeal is directed at geomatics specialists for their clearly recognized skills and qualities in terms of data. Above all, however, it is for their ability to deal with multiple themes, to engage in dialogue with diverse professional cultures, to transform data into information, to develop simple and expressive tools for accessing this information and to build and facilitate partnerships around data. All of these activities demonstrate that the primary purpose of a GIS is to contribute to a systemic approach within a structure and, in this case, within a territory.

5.6. Conclusion: toward a convergence between geonumerical sciences and territorial sciences

We have attempted to show that the vision of a "territorial trickle-down" effect of the methodological achievements of "classical" GIS is not an option, or at least that it has reached its limits by leaving a very significant portion of territories on the sidelines – on a surface area basis at least. From a conceptual and operational point of view, the idea of a "miniature GIS" does not make sense: once you start setting up and managing business databases, a dynamic is created that leads to the integration of new data and the need to maintain it. This is both a Sisyphean task and a bottomless pit, because the more specific and accurate the data are, the more it measures constantly changing real-world conditions, and the more we are led to describe things in detail, particularly in the environmental field. This then requires an increase in human and material resources. However, in terms of resources, we have seen the gap that exists in absolute and relative terms between high- and low-density areas. Furthermore, as has been pointed out, when it comes to geographic data, the area concerned is an additional requirement. It is particularly obvious in natural and rural areas where issues such as water resources, biodiversity and natural hazards become information challenges of the same level of importance as those in urban areas.

The question then becomes: What should be done? Should we simply accept this irremediable information gap between regions? Or should we consider and experiment with other ways of designing, producing, and using local spatial knowledge?

Since the 1990s and the rather laborious development of GIS, technologies have changed and the relationship with technology has been transformed, as have territories themselves. History can and must therefore be written differently. Today, we would not ask a government to start by equipping its territory with telephone lines before moving onto mobile telephony. In the same way, certain steps can be skipped to allow the diversity of territories to equip themselves with self-knowledge systems that meet both their needs and their possibilities based on the principle that there is no single model of knowledge. GIS have essentially been built incrementally, layer by layer, theme by theme, based on common reference systems. However, integrating them into a single, common data model is becoming more and more difficult, resulting in approaches that are increasingly effective but segmented. As Major and Golay (Roche and Caron 2009) pointed out in 2009, "given the diversity of cognitive models [in various professions] and the different nature of information systems, there cannot be a single unified GIS". Although large organizations have been able to cope with this situation over the years, the analysis of the CNFPT e-community shows that these GIS are very often integrated into *[data and] geographic information departments*. The term "system" no longer appears in their titles, suggesting that the uses of geographic information include GIS but are not limited to it, or at least not anymore. This amounts to recognizing and acknowledging the fact that geographic information, including in its digital form, can exist outside of GIS and even outside of geomatics.

This observation and development open up exciting avenues for research and experimentation, starting with how knowledge of the territory is conceived and mobilized by stakeholders. The aim is then to move in a targeted and "frugal" manner toward data, contrary to what has been practiced in the development of GIS. Ultimately, it is a question of laying the theoretical and practical foundations for a new profession that, for want of a better term, could be described as "geo/maieutician" rather than geo/matician. The difference then lies more in the order and manner in which skills are implemented than in their very nature. It is no longer a question of developing a "report" based on standardized but available data (and vice versa) without considering the effects, or lack thereof, on the actors involved. The primary objective is to help them develop a *shared vision* (in both the sense of observation and future prospects) by mobilizing all available resources to produce, express, formalize and ultimately share knowledge. In doing so, it is important to consider the actors as stakeholders in the process of producing understanding and self-knowledge by and for the territory. From this perspective, geographic data and the associated analysis and visualization tools clearly have a role to play, but in symbiosis with other angles or forms of understanding the "real world".

From this perspective, "small territories" have particular advantages. First, the smaller size of the structures generally makes the relationship between elected officials and services easier than in large territorial technostructures. For the *geomaieutician*, it is then possible, by taking advantage of the available data and tools, to show a representation of the territory that allows decision-makers to "find their way around" in all four senses of the term: to get their bearings, to find traces of their concerns, to share the same vision of the territory and to find their place in the methodology implemented. Such an approach requires accepting the fact that accurate or up-to-date data may not yet exist for a given theme, but this revealed absence can actually stimulate discussion and validation of its interest and, therefore, its necessity by stakeholders. It also requires the *geomaieutician* to be present and active in working or public meetings that use geographic information to understand both the requests and what "works" or does not work (in the sense of producing an effect) in the analyses or representations used, a practice that is uncommon in the daily work of geomaticians (Catteau 2023).

Another lever lies in a triple advantage. While small communities have little capacity, and sometimes little interest, in territorial data, they have nevertheless been highly mobilized around digital development of the territory. At the same time, their inhabitants are increasingly familiar with digital and even geodigital tools. Furthermore, to compensate for the limitations of public action, low-density areas are often characterized by the existence of multiple associations (personal services, socio-educational, environmental, etc.) that are aware of the diverse realities on the ground.

The existence and promotion of a "territorial information platform" can help to initiate, assist and facilitate mechanisms for collecting information on the ground and from stakeholders, instead of the professional data that the local authority is not necessarily able to produce or to order. This is not necessarily about data, but rather a set of expressions of territorial knowledge of a very diverse nature, integrated into a common digital environment, whose mobilization in situ and in vivo is more a matter of culture than of geomatics technology.

Generally speaking, without questioning what has been achieved with the development of GIS and "traditional" geomatics, it is important to recognize the need for "disruption" in areas where the same model cannot be replicated, even on a small scale. This does not necessarily mean using "low-tech GIS" since various technologies can be used, including geomatics tools, provided that the conceptual and material constraints of setting up an information system are not imposed as a prerequisite. In any case, it can be considered the responsibility of both the geomatics scientific community and the territorial sciences to ensure that a significant part of the territories does not remain, to paraphrase Noucher (2023), not in the "white" (because there is still reference data) but in the "gray of the maps", at the risk of contributing to accentuating disparities, or even territorial divides that are already largely multifactorial.

5.7. Acronyms

ADFC	*Association des Communautés de France, aujourd'hui France Intercommunalité* (Association of French Communities, now France Intercommunalité)
ADGCF	*Association des Directeurs Généraux des Communautés de France* (Association of Chief Executives of French Communities)
AMR	*Association des Maires de France* (Association of Mayors of France)
ANCT	*Agence Nationale de Cohésion des Territoires* (National Agency for Territorial Cohesion)
ARF	*Association des Régions de France* (Association of French Regions)
AFIGEO	*Association française pour l'information géographique* (French Association for Geographic Information)
BAN	*Base Adresses Nationale* (National Address Database)
BD Parcellaire	*Données cadastrales du PCI retravaillées par l'IGN pour une compatibilité cartographique* (PCI cadastral data reworked by the IGN for cartographic compatibility)
BD Topo	*Base de données à grande échelle des éléments du territoire français et de ses infrastructures* (Large-scale database of elements of French territory and its infrastructure)
BOAMP	*Bulletin Officiel des Annonces des Marchés Publics* (Official Bulletin of Public Procurement Notices)
C.A.	*Communauté d'agglomération* (Metropolitan community)
C.U.	*Communauté urbaine* (Urban community)
CNFPT	*Centre National de la Fonction Publique Territoriale* (National Center for Local Government)

ComCom.	*Communauté de communes* (Community of municipalities)
DGFiP	*Direction Générale des Finances Publiques* (Directorate General of Public Finances)
DGI	*Direction Générale des Impôts* (Internal Revenue Service)
DOM	*Départements d'Outre-Mer* (Overseas Departments)
EDIGEO	*norme d'Échange de Données Informatisées dans le domaine de l'information GEOgraphique* (Electronic Data Interchange standard in the field of GEOgraphic information)
ENEDIS	*Électricité Réseau Distribution France* (Électricité Réseau Distribution France)
EPCI	*Etablissement Public de Coopération Intercommunale* (Public Institution for Intermunicipal Cooperation)
ETALAB	*Administration publique française visant à améliorer l'action publique grâce aux données* (French public administration aiming to improve public action through data)
GAFAM	Google, Apple, Facebook, Amazon and Microsoft
GRDF	*Gaz Réseau Distribution France* (Gas Network Distribution France)
IDS	*Infrastructure de Données Spatiale* (Spatial Data Infrastructure)
IGN	*Institut Géographique National* (National Geographic Institute)
INSEE	*Institut National de la Statistique et des Etudes Economiques* (National Institute of Statistics and Economic Studies)
INSPIRE	INfrastructure for SPatial InfoRmation in the European community (European directive)
MAGIS	*Méthodes et Applications pour la Géomatique et l'Information Spatiale (réseau de recherche)* (Methods and Applications for Geomatics and Spatial Information (research network))

PCRS *Plan de Corps de Rue Simplifié* (Simplified Street Plan)

RGE *Référentiel à Grande Echelle* (Large-Scale Reference System)

SDIS *Service Départemental d'Incendie et de Secours* (Departmental Fire and Rescue Service)

SIT *Système d'Information Territorial* (Territorial Information System)

5.8. References

Beckouche, P., Grasland, C., Guérin-Pace, F., Moisseron, J.Y (eds) (2012). *Fonder les sciences du territoire*. Karthala, Paris.

Catteau, S. (2023). Bulles de dialogue et cellules raster. Spatialiser les fonctions et le fonctionnement des zones humides par SIG dans le bassin Rhône-Méditerranée Corse. PhD Thesis, Université Bordeaux Montaigne, Bordeaux.

Faure-Muntian, V. (2018). Les données géographiques souveraines. Government Report, Ministère de la Transition écologique et solidaire, Paris [Online]. Available at: www.vie-publique.fr/rapport/37550-les-donnees-geographiques-souveraines-rapport-au-gouvernement.

Feyt, G. (2002). La couverture cartographique nationale à l'épreuve des technologies. In *Ces territorialités qui se dessinent*, Debarbieux, B. and Vanier, M. (eds). La Tour d'Aigues, L'Aube/DATAR, Paris.

Feyt, G. (2007). Redistribution des pouvoirs, redistribution des cartes : la connaissance des territoires, enjeu inédit de l'action publique ? In *Les politiques publiques à l'épreuve de l'action locale*, Negrier, E. and Faure, A. (eds). L'Harmattan, Paris.

Feyt, G. and Noucher, M. (2014). La gouvernance informationnelle, outil et enjeu stratégiques des recompositions territoriales : vers l'émergence de nouveaux référentiels géographiques ? Report [Online]. Available at : https://shs.hal.science/halshs-00987788v1.

Hennion, C., Altounian, M., Monthubert, B. (2023). Rapport de la mission Data et territoires. Report, Ministère de la Transformation et de la Fonction publiques, Paris [Online]. Available at: https://www.vie-publique.fr/rapport/291763-rapport-de-la-mission-data-et-territoires.

Joliveau, T. (2004). Géomatique et gestion environnementale du territoire. Recherches sur un usage géographique des SIG. Thesis, Université de Rouen, Rouen.

Joliveau, T. (2007). Echelle 1:1 et représentation grandeur nature. Monde géonumérique [Online]. Available at: https://mondegeonumerique.wordpress.com/2007/04/30/echelle-11-et-representation-grandeur-nature/.

Lengage, G. (1999). Les perspectives d'évolution de l'information géographique et les conséquences pour l'IGN. Report, Paris [Online]. Available at: https://www.vie-publique.fr/rapport/25688-les-perspectives-devolution-de-linformation-geographique-et-les-conseq.

Major, W. (1999). Approche de la concertation territoriale par l'analyse systémique et l'analyse lexicale du discours des acteurs. Perspectives d'application aux systèmes d'information géographique. PhD Thesis, École Polytechnique Fédérale de Lausanne.

Mandelkern, D. and Du Marais (1999). Diffusion des données publiques et révolution numérique. Rapport dans le cadre du Programme d'Action du Gouvernement pour la Société de l'information (PAGSI). Report, Commissariat général du Plan, Paris [Online]. Available at: https://www.vie-publique.fr/rapport/25601-diffusion-des-donnees-publiques-et-revolution-numerique-rapport.

Mericskay, B. (2007). Les technologies de l'information géographie dans les collectivités territoriales : la géomatique intercommunale à travers l'exemple des Communautés d'Agglomérations du Muretain et du Sicoval. Master's Thesis, Université de Toulouse le Mirail.

Moine, A. (2006). Le territoire comme un système complexe : un concept opératoire pour l'aménagement et la géographie. *L'Espace géographique*, 2006/2, 115–132.

Noucher, M. (2012). La donnée géographique aux frontières des organisations : des réseaux géomatiques à l'intelligence territoriale. *Revue Internationale de Géomatique* 22(1), 7–30.

Noucher, M. (2023). *Blancs des cartes et boîtes noires algorithmiques*. CNRS Éditions, Paris.

Pornon, H. (1998). *Systèmes d'information géographique, pouvoirs et organisations : géomatique et stratégies d'acteurs*, L'Harmattan, Paris.

Roche, S. and Caron, C. (eds) (2009). *Organizational Facets of GIS.* ISTE Ltd, London and Wiley, Hoboken, NJ.

6

On the Intersection of Geomatics and Digital Humanities

Juliette MOREL[1] and Baptiste HAUTDIDIER[2]

[1]*Lab'Urba, Université Paris-Est Créteil, France*

[2]*UR ETTIS, Inrae Bordeaux, France*

6.1. Introduction

Geomatics and digital humanities: where do they intersect? To begin answering this question, let us describe the disciplinary landscape. Consider first the digital humanities: they result from the encounters of the humanities with digital technology, broadly defined here as anything that can be processed by a computer and therefore translated into algorithms and numerical data. Among the various forms of digital humanities, we can start with the questions raised by those focusing on digital technology itself: What are the imaginaries associated with digital technology? What place does digital technology occupy in arts and the imagination? What are its sociocultural consequences? A second perspective, with an increasing prevalence, considers digital humanities via a reliance on digital technology as a tool or an integration into research methods. According to this definition, which will be predominant in this chapter, it might be tempting to reposition geography from the social sciences to the humanities. Similarly, geomatics, broadly defined as the use of information technology to process spatial data and information, could then be considered a branch of digital humanities. However, integrating geomatics into digital humanities would mean ignoring the distinct histories of two academic fields with parallel development paths. The research efforts that are self-described as part

Geospatial Technologies and Society,
coordinated by Thierry JOLIVEAU and Matthieu NOUCHER. © ISTE Ltd 2026.

of digital humanities are mainly in the fields of history, archaeology, literature, linguistics, art history, design, philosophy and documentary sciences. In contrast, the geomatics community, often referred to by the more generic terms of geographic information sciences, tends to lean toward quantitative and modeling methods, or even computer science, on the one hand, and land and environmental management sciences, on the other hand. In fact, while academic geomatics has its research networks, annual conferences, specific editorial spaces and university courses, digital humanities have their own[1].

These two worlds rarely overlap at the institutional level. In research practice, however, bridges between the two are frequent and long standing. Since the 1960s, archaeology has been using geographic information systems (GIS) to manage and process its excavation records (see section 6.2.1), and the discipline of history quickly followed suit by constructing multiple historical atlases and GIS. The structuring of a "spatialized digital humanities" field from the 2010s onwards results from the prior influence of the "spatial turn", which emerged in the social sciences and humanities 30 years earlier, affirming space as a central dimension of analysis for human and cultural phenomena. These spatialized digital humanities aim to explore the spatial dimension of historical, linguistic, literary, artistic and other phenomena. They encompass a wide variety of initiatives that share a common use of digital tools, particularly those derived from geomatics (GIS, spatial analysis and geovisualization), but not exclusively. The field is thus marked by heavy investment in natural language processing (NLP), the promise of automation associated with artificial intelligence, and attempts to contribute to the Web of Data – as an effort to structure and interrelate data for better semantic exploitation (Brando et al. 2021). Numerous editorial milestones have helped to structure the field, both in English-speaking spheres (Presner and Shepard 2015; Bodenhamer et al. 2010; Murrieta-Flores and Martins 2019) and French-speaking spheres (Brando et al. 2021) – a main focus in the remainder of this chapter. In France, another major factor is emerging around the role played in humanities and social sciences laboratories by professions labeled as Information Technology (IT) support, specifically geomatics, computer science and data processing and analysis. Dedicated specialists, namely research and design engineers, are very often the driving force behind a collaboration that requires meticulous technical work and conditions its success, but sometimes also its failure. Although it is difficult to access this information, it seems crucial to explore the encounter between geomatics and digital humanities in the practical dimensions of its implementation. Who works

1 Such as the European institutions DARIAH and CLARIN, the French association Humanistica or TGIR Huma-Num; the annual DH conferences of ADHO, THATCamp; or the journals Digital Humanities and Journal of Digital Humanities.

in this field and how? In what institutional contexts? Following what professional cultures? What dialogues between disciplines and professions are at work? How do projects evolve? How are they funded? These conditions are decisive for the success or failure of such research.

Despite longstanding, abundant and structured borrowing and exchanges, many initiatives at the intersection of geomatics and digital humanities still bear the hallmarks of interdisciplinary encounters. They bring together different epistemologies, methodologies, practices and professional cultures, thereby enabling useful steps forward for both disciplines. However, while digital humanities generally recognize the contribution of geomatics to their spatialized branch, the reverse is probably not as true. The practices of digital humanities are nevertheless rich in lessons: the geolocated information handled by the humanities is generally questionable, being rarely first-hand data referring to a tangible reality about which it is possible to make accurate and objective observations (Presner and Shepard 2015). De facto, researchers in digital humanities often take a critical (or at least reflective) stance. Far from fantasies of objectifying the contemporary world, the reality sought by medievalists, archaeologists or literature specialists may be distant and distorted, or even belong to an imaginary world or artistic universe. In this respect, the extraction of information, its digital formalization, localization, storage, processing, interrogation and visualization raise specific questions. These include the rendition of the specificity of the media enabling the access to knowledge (the idea of mediality), the incompleteness and imperfection of data; the heterogeneity of sources, formats, precision and granularity; and the mixed status of real, past, imaginary and fictional information. Of course, geographic information sciences did not wait for digital humanities to explore these types of problems, but the latter provide key elements for thinking about the roles of geographic information and geovisualization in contemporary digital society.

To explore the variety of actual and potential encounters between geomatics and digital humanities, we propose breaking down the terms of the subject and considering the relationships between space, humanities and digital technology according to three preferred movements. We will thus explore:

– *The need for spatialization in the digital humanities.* In line with the spatial turn experienced in social sciences, expectations have been shifted to the humanities: gaining a better understanding of archaeological records, a text or corpus or a historical period through its spatial dimension, often by transposing the off-the-shelf digital tools of geomatics (see section 6.2).

– *The digitization of spaces, places and spatialities described in the humanities.* These encounters are more midway between the two disciplines, as they aim to propose ad hoc tools or models to address the difficulties of spatiotemporal formalization or interoperability posed specifically by the objects, sources and questions of the humanities. These developments, in turn, offer potential for more "traditional" geomatics (see section 6.3).

– *Attempts for further humanization of the digitized space of GIS,* or how, based on experiments conducted in particular in the digital humanities, it is possible to describe, in geomatics and digitally, a space that is more historicized, imaginary/imagined, inclusive, sensitive, complex, "thick" or "deep" (see section 6.4).

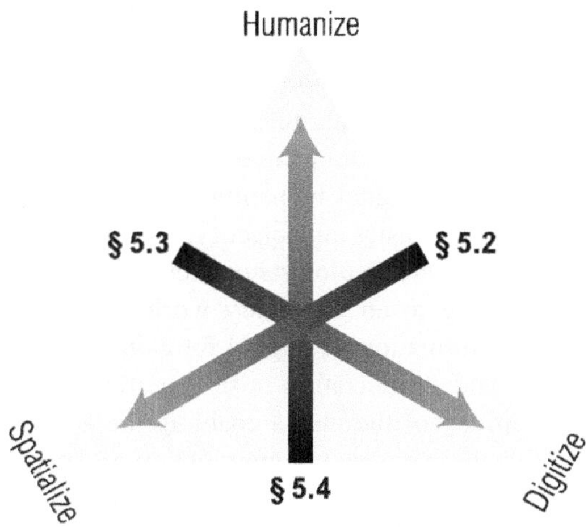

Figure 6.1. *Between space, digital technology and the humanities, three movements of encounter*

NOTE.– Finally, we should point out one bias in the presentation: the prominence given to research projects conducted in French-speaking universities. While some leading English-language projects will also be cited, the dominance of academic examples reflects, at least in France, the reality of their deployment.

6.2. Digitally spatializing the humanities

The first approach is also the oldest: it involves taking into account the spatial dimension of the phenomena, works and objects studied in digital humanities, thus spatializing digital humanities, or more precisely, digitally spatializing the humanities. The concomitance of digitization and spatialization is common to the primarily pragmatic motivations that have led to the deployment of new tools and methods in spatial sciences, geography, environmental sciences and land management. Whether it be statistics, cartographic visualization, theories and equations of spatial analysis or GIS, the fundamental need is to process large volumes of geolocated data and understand spatial dynamics.

6.2.1. *Archeology and geomatics*

Archeology is undoubtedly the oldest of the digital humanities: the *Newsletter of Computer Archeology* was founded in 1965 in the Department of Anthropology at Arizona State University. Its first editorial developed the idea of the computer as a "prosthesis" useful to archaeologists for "extending" their "intellectual capacities" (Moscati 2021, pp. 12–14). However, this type of practice did not spread before the 1970s, becoming only commonplace in the 1990s (Brando et al. 2021). While this computerization has been gradual, the links between archaeology and spatialization are older, even consubstantial. The information gathered during excavations is spatial in nature, and space is a powerful factor in analysis. This is how geomatics tools were "logically" adopted by archaeologists: first GIS for creating inventories, then spatial analysis for interpreting them (Barge et al. 2004, pp. 16–17). This was followed by a very strong technicization of archaeology that fully integrated remote sensing, photogrammetry, lidar surveys and 3D modeling, for example, among its prospective tools (Brando et al. 2021, see section 6.2.2). We will see later (see section 6.3.2) that the contributions of archaeological research in terms of modeling spatiotemporal information, making it available, and disseminating it are also significant.

6.2.2. *Spatial analysis and GIS for historical, literary, cultural and heritage studies*

Another prime area where geomatics and digital humanities have intersected is the construction of historical, literary, cultural, heritage GIS, etc. By "GIS", we mean here the creation of geolocated databases that link a variety of data through spatial dimensions and explore them using spatial analysis tools or (cartographic) visualization. These GIS are most often implemented using dedicated software (QGIS, ArcGIS, etc.) or online platforms (ArcGIS Online, GEO, ad hoc web mapping applications, etc.). They can meet a variety of needs, often related: the collection of historical, literary, cultural and heritage information with a spatial and geolocatable dimension; spatial analysis; and (carto)graphic representation for the analysis, communication and dissemination of this information. While these practices are widespread in archaeology, they have been more recently imported into other humanities disciplines, becoming commonplace in history, literature, art history and heritage, and documentary sciences. This is evidenced, for example, in France by the continued vitality of the seminar *Des sources au SIG* (since 2020), organized and run jointly by engineers from the CNRS, the Dypac laboratory at the University of Versailles-Saint-Quentin and the Criham laboratory at the University of Limoges, as well as the very long list of projects combining GIS and the humanities on the *Anterotesis* website, a tentative directory of this type of project in history (DH GIS Projects 2011).

In the field of history, the model project of this type in France is undoubtedly the ALPAGE-Paris program, coordinated by Hélène Noizet from the Paris-based Lamop laboratory (Noizet et al. 2013). It "aims to establish shared working tools [between Information and Communication Sciences and Techniques and the Humanities and Social Sciences] to develop research on the Parisian urban space, in particular through a GIS comprising cadastral and historical layers" (ALPAGE n.d.). An online GIS application (often referred to as a webGIS platform) thus enables the dissemination of vector data concerning Paris from antiquity to the end of the 19th century, as well as georeferenced old maps dating from 1750 to 1880. This program is characteristic of the scale that such projects can take on, as evidenced by the age of the project (18 years in 2024), the number of collaborators involved (20 on the program's website), the number of institutional partners (21 collaborating institutions, associated laboratories, or funders) and the amount of information presented on the webGIS platform (nearly a hundred thematic vector layers and 14 layers of georeferenced historical maps). It should be noted that the authors and specific roles listed in the project documentation relate exclusively to the production of geohistorical data and not to the design or interfacing of the dynamic web GIS.

In history, we can cite another project representative of recent productions: the *Atlas Historique du Limousin* (Morel et al. 2017, 2020). This project has three objectives: to reference all ancient cartographic representations of Limousin (a former French NUTS2 administrative region), to provide an editorial space for all geohistorical studies conducted on this territory at different scales, and to enable dynamic consultation of ancient cartographic documents and spatiotemporal vectorized data describing the urban history of Limoges, the regional capital. The project was funded and carried out locally, respectively, by the former Limousin region and the University of Limoges between 2014 and 2018. Although informed by ongoing dialogue, the organization of this project is marked by a strong division of labor between history and geomatics (even though each of the geomatics specialists involved was experienced in the field of humanities data or trained in history).

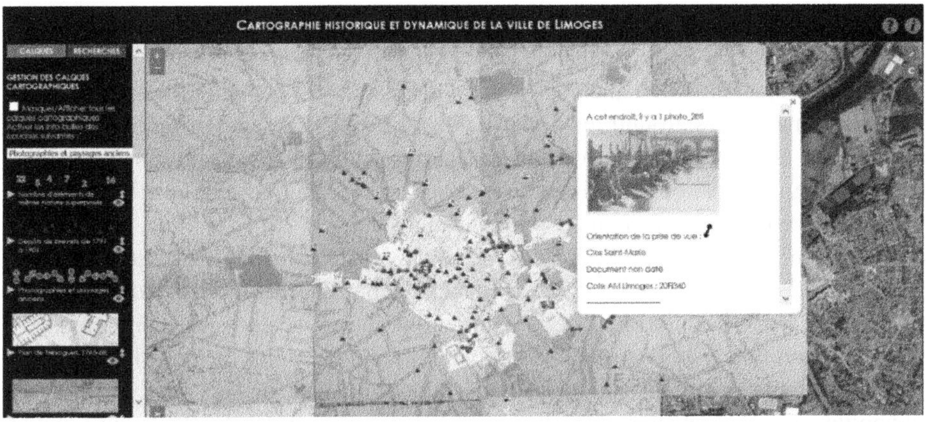

Figure 6.2. *Screenshot of the web interface for querying and mapping historical documents on contemporary Limoges, as part of the Historical Atlas of Limousin. Created by Juliette Morel, 2017. The web mapping section of the site will no longer be maintained in 2024. For a color version of this figure, see www.iste.co.uk/joliveau/geovisualization.zip*

In the field of heritage and art history, digital atlases and webGIS are legion. Their appeal often lies in the fact that they have been developed outside of academic research by associations or public cultural actors. Examples include the Atlas of Heritage and Architecture developed by the Conseil départemental de la Seine-Saint-Denis (2024) and the Atlas of Heritage (2021) published by the Ministry of Culture, which lists classified, registered and protected sites. There are many examples of simple cartographic web interfaces that locate the main heritage sites in a given area and provide a descriptive file for each one. Another example is the *Geoculture* – Limousin as seen by artists (Thuillas 2010). This associative website offered dynamic mapping of the location of works of art, comparing them with reproductions of the works themselves (mainly literary texts, photographs, paintings and sculptures) with the aim of applying the principles of Bertrand Westphal's geocriticism (2011). In the field of documentary sciences, the Gallicarte project aimed to add a dynamic web map to the BNF Gallica documentary research platform, which would have made it possible to locate the places mentioned or represented in the resources, thereby capitalizing on the spatial richness of the data contained in the Gallica catalog. Starting out as an automatic process, the geolocation of resources was opened up to collaborative features (Gioux 2018). However, the service was discontinued, with OpenStreetMap mapping of contributions ending in February 2020 (BnF Gallica 2020).

In the field of literature, the *Mapping the Lake District* project is a pioneer. Led by Ian Gregory and Sally Buchel at Lancaster University in the United Kingdom, it develops a "spatial textual analysis" using GIS features to compare the spatialities of two travel narratives around the Great Lakes of England by two different authors at the turn of the 18th and 19th centuries. This project emphasizes the importance of geolocating information from literature, taking into account the linguistic context to introduce perceptual information (Cooper and Gregory 2011; Gregory et al. n.d.). Here, we see that literary text is a unique source of information that requires special treatment. We can also note an "original" technical choice: instead of a fully integrated webGIS interface, geographic information layers and instructions for viewing and using them are made available on Google Earth. Many project examples have opted instead to provide an online literary GIS or dynamic web mapping, such as the Literary Map of NYC (n.d.) or *Mapping St Petersburg*, which locates the novels of Gogol and Dostoevsky in the city (Young and Levin n.d.).

6.2.3. *Cartographic visualization in the literary humanities*

Among these efforts for the localization of literary sources, some focus on cartographic visualization as a method of analysis and a powerful communication tool. Franco Moretti, an Italian researcher in literary studies who has worked mainly in the United States (Columbia and then Stanford), is one of the theorists behind this approach in digital humanities with his method of distant reading. He proposes considering a large corpus of works with quantitative tools enabling their processing: statistics and, above all, graphical representations. Having put this approach into practice in an analysis of the locations of 18th-century European novels (2000), he theorizes it in *Maps, Graphs, and Trees* (Moretti 2005), where he deploys these three graphical modes to highlight major patterns in literary history. Here again, digital technology and graphical design are used for practical reasons: the masses studied could not be addressed otherwise.

Other research projects take a similar approach, mapping spaces and places from large corpus of literary texts to better understand them. One example is the work of Barbara Piatti's team on the Literary Atlas of Europe (Piatti et al. 2008; LiteraturAtlas n.d.). Due to its remarkable scope (ongoing since 2006, with 15 contributors), the project has become one of the major references in the field. The project studies "places of literary action", meaning that it seeks to answer a very general question: "Where does [European] literature take place, and why there?" using three approaches. The first consists of choosing a region and mapping the "places of literary action" of the significant texts located there. This is the "mesogeography of a literary region". The resulting visual representations make it possible, for example, to identify literary sub-regions within these mesoregions. The second method, on the other hand, consists of mapping an entire work – this is the "geography of an isolated text". Finally, a third approach proposes mapping a "literary macrospace" in a much broader spirit of synthesis. The various studies use spatial analysis tools, especially to analyze location densities. Behind all this is a database, a description of its model and an interactive mapping tool enabling data to be consulted in "GIS mode" has been developed. However, this is clearly not accessible to the public, and the project is actually communicated through summary maps, more in the style of an online atlas than a GIS.

The *Mapping the Republic of Letters* program (Gregory et al. n.d.) is also a good example of the circulation of Franco Moretti's theories. It studies, mainly through case studies, the geography of authors (rather than works) of the "Republic of Letters" in the 18th century, expressing networks of correspondence and places of publication through graphs and maps.

The use of (cartographic) visualization to study literary works is justified in other cases by the need to change perspective or scale when dealing with texts that are too complex in structure. The aim is not to adopt a distant reading, but rather to focus on a work of interest. The multiple attempts to visualize James Joyce's novel *Ulysses* are the best examples of this. Vladimir Nabokov used cartographic representations in his lectures on Joyce's work, urging that "instead of perpetuating the pretentious absurdity of these Homeric, chromatic, and visceral chapters, teachers should prepare maps of Dublin, with the intertwining paths of Bloom and Stephen" (Talmey 1969). Later, Grange (2002) developed several modes of representation to understand the construction of the novel, Pericles Lewis proposed an interactive map of Dublin to help his student readers locate the narrative episodes, and Travis (2015) used *Ulysses* as one of his case studies to deploy a method of literary mapping. This same logic of mapping to better understand narrative complexity guides the geonarrative mapping experiment we conducted on the literary work of Algerian author Kateb Yacine (Morel 2019). As the structure of the work is marked by spatiotemporal and narrative breaks, as well as internal and external references that are difficult to follow, we hypothesized that overall visualizations, combining spatial, temporal and narrative dimensions, would make it possible to identify certain geonarrative logics that are too complex to be identified otherwise.

The cartographic representation of fictional and imaginary worlds is also a hot topic in literary cartography. The recartographies of Thomas More's *Utopia*, Tolkien's *Middle Earth* (of which there are many) and Julien Gracq's *Rivage des Syrtes* by Yves Lacoste show that their purpose is not to shed light on existing spaces through the lens of the fictions that inhabit them, but rather to truly understand the internal spatiality of the works. Finally, let us mention a cartographic approach whose objective was to build a bridge between reality and fiction: *(e)space & fiction*. According to Joliveau (2015), digital maps can be used to visualize the connections between real space and fictional space: the location of fictional places when they refer to a real place, the location of filming locations or real places that inspired imaginary places or the location of the "materializations" of fiction in reality (statues or commemorations, etc.). Through the possibilities for transmedia connections offered by digital technologies, particularly the web, the *(e)space & fiction* project explores a third space in Edward Soja's sense, in that it links spaces considered to be both material and mental, as well as the incessant reciprocal influence and collisions between reality and fiction (Joliveau 2015).

Many digital humanities projects have therefore adopted geomatics tools. Whether in the form of GIS in archaeology or history, online mapping platforms in the field of heritage or documentary research, or cartographic visualization in the context of literary studies, the process is always more or less the same: it involves transposing existing geomatics tools to locate phenomena and better understand their spatiality. That said, these phenomena are not quite the same as those usually studied by geography and geographic information sciences: they are worlds accessible through often tenuous material traces or often imperfect archives, works of art and literature. In fact, their particular nature reveals difficulties that in turn require the development of specific tools.

6.3. Digitizing the humanities space

A digitization phase, meaning the translation of analog information into data that a machine can store, process and communicate, is at the heart of any digital enterprise, whether it be geomatics or digital humanities. The transition of humanities corpora to digital formats poses particular challenges at the heart of digital humanities, just as geomatics has, since its inception, had to grapple with issues of digital formalization of locations, associated semantic information, metadata, etc. The combination of the two obviously presents many challenges, particularly in terms of spatiotemporal modeling. Encounters between geomatics and digital humanities have often provided opportunities for very fruitful exchanges, allowing in particular geomatics to take a more critical view of the digitization models it applies. In other words, the space described in the humanities, through archaeological traces, historical sources, works of art or fiction, is difficult to model digitally because it is rarely described in a uniform or exhaustive manner and is generally complex. How, then, can tools and models be adapted so that this space can be digitized satisfactorily?

6.3.1. *Data extraction issues*

The first difficulty lies in collecting data in the humanities and building up digital data sets large enough to enable meaningful digital processing. Extracting data from textual sources, especially manuscripts, or iconographic archives is no trivial task. Document scanning does not equate digital "reading": it is still necessary to recognize the characters, words and sentences of an ancient manuscript (a process known as transcription, or sometimes OCR), understand the meaning of certain words to extract spatial references, geolocalize ancient maps and vectorize the information, etc. These operations, which are necessary for any subsequent geonumerical exploitation, are still largely carried out by humans and are extremely time consuming. However, there are several ways to optimize them.

The first involves setting up participatory processes, in that case digital interfaces allow large numbers of people to get involved in the digitization of sources. In the realms of humanistic geographic information, some noticeable projects are more or less open and backed by citizen science processes, such as the New York Public Library's *Footprint Checker* platform, which aimed to get the general public to draw, in a gamified form, the footprint of New York buildings on old maps of the city (Wieck 2018), or *Gallicarte* in its collaborative aspect (see section 6.2.2). Other projects are more specifically aimed at networks of specialists, such as the creation of the *Collégiales* (listing medieval French collegiate churches) and *Monastères* (listing medieval French monasteries) databases, brought together in the Col&Mon program (Deflou-Leca and Massoni 2020). Both call on history specialists to provide information from their research via online interfaces accessible after identification.

A second approach to addressing the difficulty of building spatial data sets in the humanities is the automation of information extraction, particularly spatial information. This is a vast field of research in the digital humanities, particularly in relation to NLP. As for the extraction of spatial information from texts, the challenges are numerous but relatively well defined: automatically recognizing direct geographical references, most often through place names, but also – and this is where the difficulties lie – when places are referred to in relative terms ("near", "at someone's place", "three kilometers from", etc.). These steps involve both "named entity recognition" (NER) and semantic annotation aimed at disambiguating terms (or name entity linking), in this case spaces and places, to achieve accurate locations (Brando et al. 2021; McDonough et al. 2019; Moncla et al. 2016). One challenge at this semantic level lies in characterizing the locations thus extracted: What do literary sources, which are never neutral, say about them? With what connotations? According to what representations? (Cooper and Gregory 2011;

Dominguès et al. 2019; Alrahabi et al. 2021). It is easy to imagine the broader potential of these geomatics tools for integrating geographic information from new and often alternative sources. The transposition of these methods was not long in coming: for instance, the ontology developed as part of the Choucas project aims to locate victims of mountain accidents using heterogeneous sources and fuzzy spatial information, particularly textual information derived from verbal exchanges between victims and rescue services, using NLP methods (Bunel et al. 2019). In another field, the automatic vectorization of old maps is among the significant attempts to automate the extraction of information from historical sources (Odie 2017; Andrieu and Raveaux 2021; Follin et al. 2021). While the promises of a geohistory enhanced by deep learning may seem obvious, these techniques are still experimental, requiring significant data cleaning, which is always done retrospectively and manually.

6.3.2. *Data formalization issues*[2]

A second difficulty in the process of digitizing humanities data lies in the formalization of the data itself, that is, in the concrete and material form that the transition from the information contained in the analog source (text, manuscript, old map, image) to digital data will take. This difficulty is twofold for the data model: choosing precise and unique modalities (values, terms, etc.) to describe complex information, and defining an overall structure for the dataset.

The highly specific needs of archaeology and history for spatiotemporal modeling have led to deeply reconsidering the formalization of their spatiotemporal data. Building ruins, remains, artifacts, books, texts, etc.: the materials on which these two disciplines rely are multidimensional and have complex temporal, geographical, geometric and semantic characteristics. But their traces and references are often very incomplete: there may be a mismatch between the footprint of excavated remains and the entire building, the date of construction may not be known precisely, nor its uses. At the same time, this building may have undergone several phases of construction, and therefore several geometries for several dates (themselves not precisely known), and even several changes of function. Furthermore, the nature of archaeology and history is such that it is impossible to gain a completely exhaustive understanding of a subject: archaeology will never uncover all the temples of the ancient world, but will only ever know about those whose traces have been preserved and discovered; history will never have access to textual sources describing the entire medieval world, but is limited to the sources

2 Sections 6.3.2 and 6.3.3 largely draw on elements of a presentation given at the SAGEO 2023 conference at Laval University in Quebec (Morel and Hautdidier 2023).

that have been preserved. There are therefore many places and spaces in the medieval world that are attested to, or even described, by archaeological or historical evidence, but whose location is still impossible to determine due to a lack of remains. Uncertainty about the correspondence between excavations and documentary references is also common (Elliott 2018; du Mouza and Lamasse 2022). Finally, this knowledge is as fragmented as it is evolving, hence the prevalence in archaeology of the idea that modeling focuses on a "state of knowledge" (Elliott 2018).

Historical and archaeological disciplines, which are ontologically marked by significant challenges in terms of digitization, have thus proposed to challenge the model of GIS, at least in the narrow sense of GIS software. The argument is that these systems impose excessive modeling constraints, particularly due to the differentiation of information into hermetic layers of information with distinct geometries (points, lines or polygons), granularities and themes. This idea can be found, for example, in the *un-GIS* proposal made as part of the Pleiades reference project, which aims to bring together archaeological and historical knowledge about ancient sites (Bagnall et al. 2006); in the data structure of the Col&Mon project, which aimed to merge several existing databases on medieval religious establishments and study their spatiotemporal settlement characteristics through a conceptual "faceted" model (Deflou-Leca and Massoni 2020); or in the OH_FET (for historical object/function/space/time) conceptual data model developed by archaeologists at the University of Tours (Rodier and Saligny 2010). All these models are based on the dual principle of discrete information processing through atomization and description using a space/time/theme triplet (Peuquet 1994). These choices enable the following:

– The disambiguation of spatiotemporal empirical traces or documentary references from the geo-historical objects they describe. Specific sources of information become the central entities of the database, and the objects reconstructed through analysis emerge from relationships between these elements. Interpretation thus regains its rightful place as an abstraction reconstructed from concrete observations.

– The integration of spatiotemporal data exactly as collected, without generalization. This makes it possible to combine data with heterogeneous spatial or temporal granularities and to reconstruct spatialities or temporalities by inheritance between related elements.

– The processing of the entirety of the database in queries, thanks to a symmetrical treatment of its dimensions (Peuquet 1994) and a tolerance for incompleteness, particularly spatial incompleteness.

The principles cited allow for a certain agnosticism regarding the implementation of databases, depending on the technologies mastered and the intended uses: while there is growing use of graph-based data modeling (in the case of Pleiades) and, more generally, the Semantic Web, relational databases are also present (in the case of Col&Mon or the Historical Atlas of Limousin, for example).

These models echo other areas in which similar questions arise, but where solutions have undoubtedly not gone as far. In literature, for example, uncertainty about the location of a place mentioned is common: because it is unknown (too small, unlisted, designated by a local place name, etc.), vague (designated in a relative, imprecise manner), or imaginary. In all cases, places in a work with similar symbolic or poetic status may have different scalar granularities or geometries, which sometimes makes their digital formalization within a traditional GIS difficult. In Kateb Yacine's novel *Nedjma* (see section 6.2.3), spaces with different statuses coexist: the cartographic form of the polygon, representing indeterminate and global territorial entities, is very important, and certain spaces such as colonized Algeria, ancestral regions, or metropolitan France are treated as such in the text. Other spaces are very specific and real. Others are indeterminate, such as "the city of X" or "the construction site". Finally, others are imaginary, such as Nadhor, a mythical place of birth for the characters. If all these spaces, including those that cannot be located, are equally important in understanding the geography of the work, how can they be integrated equally to produce (carto)graphic treatments and representations that shed light on the geography of the work? In hindsight, we can consider that the atomized, three-dimensional modeling of archaeology, or the choice made by the *Literary Atlas of Europe* to situate literary spaces on a "scale of reference and uncertainty" (Reuscher et al. 2012) would have been of great interest for structuring the database.

These principles and questions, which focus as much on formalizing and structuring data as on taking uncertainty and incompleteness into account, relax the constraints of GIS structuring and allow for the reintroduction of a constructivist conception of knowledge into spatiotemporal modeling.

6.3.3. *Issues relating to data availability and dissemination*

A third question raised by the digitization process in the digital humanities/geomatics encounter has to do with data dissemination. This question does not only arise at the end of the process: it is often the *raison d'être* of projects and determines all the technical digitization choices involved in the process.

In the context of archives and libraries, numerous initiatives aim specifically to enable the consultation of geolocated documentary resources. Two types of projects can be distinguished: those aiming to use the geographical information already present in library catalogs, and thus to geolocate the catalogs, such as the above-mentioned *Gallicarte*; and those aiming to georeference, visualize, vectorize and make available for download collections of old maps. A long-standing collaboration between the French national institute of cartography (IGN) and the School for Advanced Studies in the Social Sciences (EHESS), the *GeoHistoricalData* project (Perret et al. 2015) is part of this "digitizing" ambition. It was inspired by the observation that there was a lack of cumulative effort in the tedious work of digitizing old maps carried out in the context of scattered specific research projects. The platform provides georeferenced versions of Cassini's maps (18th century) and plans of Paris dating from 1728 to 1888, as well as downloadable layers resulting from selective vectorization (road networks, cities, forests, etc.). Another pioneering example is the *Cartomundi* digital map library (Arnaud 2008), whose aim of promoting a cartographic heritage initially focused on the Mediterranean has been particularly well served by the central place given to composite maps in the presentation and exploration of sources. We should also mention *Navigae*, which is based on the collections of French geography laboratories belonging to the ImaGEO consortium. Its originality lies in its symmetrical treatment of maps and photographs, which are treated as geographical images within a common web interface (Kilouchi 2023).

Another field of spatialized digital humanities focuses on this goal of knowledge circulation: geographical indexes, or more commonly gazetteers, which are directories of places associated with digital locations. *GeoNames* has established itself as the leading gazetteer for place names, but others orbit around it, with an inspiring network structure: for instance, *Pleiades*, dedicated to ancient places, draws on *GeoNames* for the location of current place names, on the *PeriodO* thesaurus for historical periods and on the *Getty Art and Architecture Thesaurus* for descriptions of artifacts (Elliott 2018). These directories are inseparable from the Web of Data in which they are embedded. They feed into interconnections whose potential for semantic enrichment is staggering, but which are only made possible upstream by long-term work to harmonize and standardize data description models, notably under the leadership of the Open Geospatial Consortium. In addition to the requirement to describe the underlying characteristics of the information to be formalized as accurately as possible (see section 6.3.2), there is also a need to adopt common ontologies. In this regard, we can mention, for example, the work of the Dataforhistory team led by Francesco Beretta at the Larhra laboratory in Lyon to enrich conceptual models specific to cultural heritage (CIDOC-CRM in this case) to adapt them to historical data and, in particular, to their spatial dimension. For digital humanities projects of this scale, data harmonization and openness are key conditions for success. Given the immensity of the task of digitizing humanities corpora and the ambitious goals of opening up research data, two developments are worth noting. On the one hand, the application of FAIR principles, which advocate making data findable, accessible, interoperable and reusable, has led to significant improvements in terms of accessibility and therefore circulation. On the other hand, the integration of data into the web promises to enrich information and create large cross-disciplinary databases (in terms of institutions, research teams and disciplines).

This opening up of geographic data deployed in the digital humanities offers immense potential for the historicization of geographic information sciences and the enrichment of geographic information, particularly its metadata. It is largely linked to an ethos of openness largely underpinning the work culture of librarians and the disciplines of documentary sciences. It is also enabled and conditioned by a technical culture closely linked to the Web of Data, as well as by the digitization and georeferencing campaigns of cartographic archives carried out by universities, archives and libraries.

6.3.4. *What promises does big data hold for the humanities and social sciences?*

One concept, or at least one keyword, may replace another: since the 2010s, many efforts have focused on "time machine" projects, with the city of Venice serving as an early showcase. Having become an associative network after being structured as part of a vast European project, the Time Machine (2024) organization has announced its intention to "develop the *Big Data of the Past*, creating a huge distributed digital information system mapping the European social, cultural and geographical evolution across times. This large-scale digitization and computing infrastructure will enable Europe to turn its long history, as well as its multilingualism and interculturalism, into a living social and economic resource".

If we compare the scale of such a promising economy with all the challenges listed so far, we can gauge the inherent weaknesses and ambitions of these initiatives. Carried out in a European research area receptive to the issue of heritage promotion, the proposal involves the prospect of promoting (geo)historical data as a new frontier to be explored. These ambitions are not limited to the expansion of knowledge resulting from the industrialization of digitization. They also involve forward- and backward-looking simulations which, using materials including landscape developments and 3D models, are close to the form and ambition of diachronic digital twins, as virtual models of physical objects (in this case, urban fabrics). These authors can be distinguished by their call for urgent investment in an archival dark matter that would only be accessible by machine (Di Lenardo and

Kaplan 2023). The same authors may also take a more critical stance on the threats associated with mass data collection, particularly in terms of surveillance (Goguey 2022). The promoters of such projects are therefore treading a rather unique line, with success that is still difficult to assess today.

6.4. Humanizing the digital space

These digitization attempts support a final mode of encounter between digital humanities and geomatics, which stems more from geography and regional sciences. This is a trend in geographic information sciences, critical or heterodox one might say, toward the "humanization" or "rehumanization" of geodigital space as it is conceived and formalized in geographic information sciences. The aim is therefore to draw on the tools and sources of the humanities to add complexity to digital descriptions of lived spaces.

6.4.1. *Complexifying digital space: historicizing, critiquing, raising awareness and narrating space in GIS*

An initial movement toward "rehumanization" is based on the massive digitization of old cartographic sources, as mentioned above. This approach allows for the historicization of region-specific information systems, as well as the analyses that they enable. There is a noticeable trend toward the publication of old maps on the general webGIS platforms of local authorities. The IGN's Geoportal provides access to Cassini maps, the General Staff map (*Carte d'État-Major*, 1820–1866), the 1950 IGN topographic map and aerial photographs taken between 1950 and 1965 covering the entire country. The *Géoportail93* of the Seine-Saint-Denis *département* (NUTS3 region) has also integrated 28 geolocated maps providing a diachronic view, representing all or part of the territory between 1731 and 2012. Much remains to be done, particularly with regard to integrating this information in vector format so that it can be taken into account in spatial analyses, which form the basis for planning decisions, but the trend itself is remarkable.

Behind the challenge of historicizing territories and spatial phenomena lies the more general question of the increasing complexity of GIS. It is described in critical literature as highly caricatural and standardized (Goodchild 2015), marked by technical "hypertrophy" to the detriment of the specific characteristics of the space studied (Desbois 2015), or reflecting an overarching and dominant perspective (Pavlovskaya and Martin 2007). A critical current in geomatics is based on these observations and proposes *Doing* critical GIS (Mahmoudi and Shelton 2022). Sometimes, this involves implementing methodological approaches similar to those developed in spatialized digital humanities, or which could greatly benefit from them (although we should point out here that digital humanities do not always adopt a critical approach, far from it). There are various ways of doing this: the critical geomatics community advocates a more reflexive and situated epistemology, placing spatial and social justice at the heart of its concerns when choosing which data or tools to use. The need to take a step back from geomatics methodologies and document the processes put in place in great detail has also been argued (Kelly and Bosse 2022). More fundamentally, it is also a question of changing the subject or point of view by focusing on themes that are often overlooked by quantitative GIS data, and giving a voice and a hand to geomaticians with less "traditional" profiles: indigenous people, dominated groups, gendered groups, racialized groups, queer people, etc. It may also involve reintegrating heterogeneous, multiple and mixed geographic information, rebalancing or even countering the quantitative, distanced and overarching description of the world with more qualitative interpretations – situated at a human level – of the territories and the social dynamics that unfold there.

The "qualitative GIS" movement, drawing from mixed methods, has been bringing together experiments of this type in the English-speaking research world for several years (Cope and Elwood 2009). The term "qualitative" implies providing "rich contextual details on material or social situations". It means taking on a dimension of phenomena that is not only descriptive but also interpretive, or aiming to "understand situated and negotiated knowledge". We can see the paradigm shift from traditional GIS, usually self-described as neutral and descriptive. Although it does not directly define itself as such, one example that we find representative is the *Anti-Eviction Mapping Project* (AEMP). This mapping project has been underway since 2013 and is the work of a group campaigning for the defense of people evicted from their homes in the San Francisco Bay Area, as well as providing geographic information on housing and property. This activist group, led by Erin McElroy, has developed a mixed qualitative mapping approach. It acknowledges that high barriers to accessing data and tools lead to "dependence on technocapitalist data and infrastructure" but that organizations producing alternative housing data do not share its goal of spatial justice. The approach combines ethnographic documentation of evictions with dissemination via dynamic maps, combining traditional Euclidean localization, quantification of evictions, categorization of types and conditions of evictions, and testimonials from victims in the form of audio or video clips (McElroy 2022; Zwer 2023). The inherent complexity of such systems is evident due to the diversity and heterogeneity of the media, voices, formats, sources and interpretations they bring together. Finally, we can see the value of drawing inspiration from the tools and processes of the digital humanities described above, particularly the potential for multimedia publishing and documentary contextualization offered by the Web of Data.

In the same critical vein, some projects aim to refocus GIS on the spatiotemporal and discursive dimension through which space is known or understood. The goal is to account for two inevitable mediations in our understanding of our relationship to space: practices and discourses. A significant milestone in this regard was Mei-Po Kwan's cartographic work on the spatial practices of Muslim women following the attacks of September 11, 2001, in an article with the evocative title: *From oral histories to visual narratives: Re-presenting the post-September 11 experiences of Muslim women in the USA* (Kwan 2008).

"Nada's life path on a typical weekday before September 11" "Nada's life path several days after September 11" "Nada's life path several weeks after September 11"

Figure 6.3. *Spatiotemporal maps from Mei-Po Kwan's article (2008), which show the influence of the visual language of Time Geography (Hägerstrand 1970). For a color version of this figure, see www.iste.co.uk/joliveau/geovisualization.zip*

The formalization and representation of how space is narrated or reported, rather than space itself, is also at the heart of *Atlasciné*, developed around Sébastien Caquard at Concordia University in Montreal (Geomedia lab 2018). This animated mapping tool makes it possible to represent the narrative importance of places in films, measured by the cumulative film time devoted to each place, whether through an explicit shot or an evocation. Another interesting aspect of this project is its transposability, since the tool developed has been reused to show the space of the narratives of survivors of the Rwandan genocide. In this case, it is absolutely necessary to inform the space by adhering as closely as possible to the narrative that is being told; the device that stages the unfolding of the narrative through the juxtaposition of a map, a video and text achieves this. These issues of annotation and mapping are particularly acute in the case of mass violence, for which only fragmentary sources remain. Several recent efforts related to the history of the Holocaust have thus used digital mapping to reconstruct and retell the trajectories of deportation (Le Noc 2021; Cousin 2023). To best reflect the specificity of the spatialities emerging from survivors' testimonies, Westerveld and Knowles (2021) have argued for the need to break free from the limitations of traditional GIS, favoring a topological logic capable of loosening the topographical constraints (i.e. "the grid").

6.4.2. *Deep mapping*

Deep maps are another emblematic example of a desire for radical complexity in our relationship with space and cartographic devices, as they promote an interweaving of narratives, media and perspectives. Bodenhamer and his colleagues (2015, p. 3) define these so-called deep maps as a "finely detailed, multimedia depiction of a place and the people, animals, and objects that exist within it and are thus inseparable from the contours and rhythms of everyday life". Characterized by layering, multiple sources and a bottom-up approach that emphasizes human experience and living things, these deep maps nonetheless cover quite diverse realities, both in terms of practices and disciplinary legacies.

The term "deep map" was coined by writer William Least Heat-Moon, in *PrairyErth: a deep map* (1991). This 600-page book is a vertical writing project devoted to the intensive exploration of a depopulated county in Kansas. The text is methodically organized around a survey of the landscapes and history of the quadrants of the topographic maps drawn up by the United States Geological Survey on the territory. Heat-Moon evokes human and non-human protagonists with humor and a form of humility that borders on the religious. However, while maps punctuate each of the 12 chapters, they are not the heart of a work that remains almost exclusively literary.

Another author, whose interest in maps led him to become a well-known figure in critical cartography, has also been noted as one of the pioneers of deep mapping: Denis Wood. In 1974, when asked to lead a landscape architecture workshop for students at the University of North Carolina State School of Design, he identified cartography as a valuable educational tool. He observed that these undergraduate students were unfamiliar with the usual techniques and expectations of maps, and would later be confronted with them, in a potentially impoverished form, in professional practices related to urban planning and development. He therefore felt it would be useful to challenge them with an exercise that valued creativity and the primacy of sensory experience. This is how an "experiential" mapping approach to the Boylan Heights residential neighborhood in Raleigh was built up over the course of several student promotions through an accumulation of more or less minimalist or experimental layers: hydraulic infrastructure, streetlights, fences, telephone cables, garden chimes, Halloween lanterns, autumn colors of tree leaves, radio waves, absent owners, etc. As recounted by the author (Wood 2015; 2021), the formalization of the approach has only been gradual and partly a posteriori. Initially conceived as an atlas, the project's narrative dimension became more apparent before ultimately adopting a poetic approach to cartography. Having made only partial and belated use of digital technology, the educational project came to a halt, before receiving almost fortuitous recognition in 1998, culminating in a much later publication (Wood 2010).

Despite a complex and intermittent development, these approaches have gained traction over the past 30 years. They have influenced both artists claiming to be counter-mapping (Bloom 2018) and members of the epistemic community of spatial humanities, who saw them as a way to overcome the limitations of traditional cartographic productions (Bodenhamer et al. 2015). In this movement to legitimize deep maps as a new form of activist or academic production, several intellectual shifts can be noted:

– The effort has become multidisciplinary, due to a shift in concerns and personnel toward the field of historical sciences.

–Tthe technical possibilities offered by digital tools have helped to fuel and renew the expectations on deep maps, particularly in their ability to provide immersion and sensory experience.

– Despite a clear programmatic and institutional shift, the new promoters of the approach seem to have integrated the need to maintain an open and not completely instrumental conception of deep maps, presenting them as a platform, a product, but above all a process.

Identifying five basic principles of deep maps, Bodenhamer (2021, p. 7) considers that they must be "*flexible*, inviting exploration; *user-centric*, supporting differing views; *path-traceable*, supporting narration; *open*, admitting new material; and *immersive*, evoking experience". These authors also pay close attention to the representation of spatiotemporal dynamics. Bodenhamer (2018) sees deep maps as the resurgence of a quest for Bakhtin's chronotope (1978)[3]. Corrigan considers them to be facilitators of a "genealogy of emplacement", understood as an investigation into a "historically contingent layering of structures, strategies, tactics, discipline, antidiscipline, environment, everyday practice" (Corrigan 2015, p. 63). Despite a wide disparity in forms and approaches, a certain common ground remains. While Wood's approach in his atlas can thus be understood as falling within the realm of poetics in the sense used by Jakobson (1963), who sees in this function of language an interest in the very form of the message, this approach is certainly not devoid of criticism or subversive potential. In this regard, it is noteworthy that a root

3 A concept from literary theory, reflecting the inseparable nature of spatial and temporal dimensions when expressed through narrative.

commonly claimed by members of the deep maps movement is psychogeography and the practice of *dérive* as initiated by the Situationist International. To paraphrase the formula from *The Society of the Spectacle* (1973)[4], these authors acknowledge that the world has already been mapped, but that it must now be transformed.

6.5. Conclusion

From the spatialization of digital humanities to the "humanization" of geographic information sciences, this is the arguably partial picture we can paint of actual and potential encounters between geomatics and digital humanities. Finally, we would like to return to three features that strongly influence the analysis:

– *The importance of professional and material conditions*, which are very often overlooked, in the development of projects and in the analysis that can be made of them. Although we have barely scratched the surface in this chapter, it would be important to document the work currently being done and the consequences of multidisciplinary and interprofessional dialogue on research results. Such an ambition requires specific methods of investigation and would be an interesting avenue for furthering knowledge in this field of research. The difficulty of writing the recent history of the field stems in part from the difficulty of accessing the research results themselves: due to maintenance constraints and technological developments, many online projects are subject to rapid obsolescence or even silent disappearance.

– *Structuring a field under pressure*: Faced with a proliferation of tentative initiatives, even "DIY" research projects that are scattered and often compartmentalized by discipline, the institutionalization of a highly standardized field of research is taking place, particularly in the area of spatialized digital humanities. This tension results in a significant difference in visibility between different research projects and, at times, the detrimental "reinvention" of existing methods, tools and data (which in turn legitimizes initiatives to structure the field).

– *The importance of an economy of promise* in a field made doubly "fashionable" by the labels of digital and spatial, with potential subsequent disappointments. There are recurring difficulties in establishing sufficiently large and usable volumes of data. And when they do exist, those large volumes of data may be of low semantic value. A challenge linked to the singular nature of the objects of digital humanities is the limited prospects for reproducibility in tools and methods. This ties in with the observation that there is a fragile balance between

4 "The world is already filmed. Now it is a matter of transforming it" (card 12'50"). Debord's formula is itself borrowed from Marx (*XIth thesis on Feuerbach*).

external expectations and the desires of members of the field, who are sometimes critical but more often concerned about the instrumentalization of their productions.

6.6. References

ALPAGE (n.d.). AnaLyse diachronique de l'espace urbain PArisien : approche GEomatique [Online]. Available at: https://alpage.huma-num.fr/.

Alrahabi, M., Brando, C., AlKhalil, M., Dichy, J. (2021). Paris dans les récits de voyage d'écrivains arabes : repérage, analyse sémantique et cartographie de toponymes. *Humanités Numériques,* 2021, 3.

Andrieu, D. and Raveaux, R. (2021). VECCAR – vectorisation automatique de cartes anciennes. *Interdisciplinarité(s)*, Aix-en-Provence.

Appel à communication 2023/2024. (2020). Séminaire "Des sources aux systèmes d'information géographique : des outils pour la cartographie dans les humanités numériques". [Online]. Available at: http://psig.huma-num.fr/seminaire-sources-aux-sig/.

Arnaud, J.L. (2008). A new website for cartographic resources – CartoMundi. *LIBER Quarterly*, 18, 271–275.

Atlas des patrimoines (2021). [Online]. Available at: http://atlas.patrimoines.culture.fr/atlas/trunk/.

Bakhtine, M. (1978). *Esthétique et théorie du roman*. Gallimard, Paris.

Bagnall, R., Talbert, R.J.A., Bond, S., Becker, J., Elliott, T., Gillies, S., Horne, R., McCormick, M., Rabinowitz, A., Robinson, E. et al. (2006). Pleiades: A community-built gazetteer and graph of ancient places [Online]. Available at: http://pleiades.stoa.org.

Barge, O., Rodier, X., Davtian, G., Saligny, L. (2004). L'utilisation des systèmes d'information géographique appliquée à l'archéologie française. *ArchéoSciences Revue d'Archéométrie*, 28, 15–24.

Bloom, B. (2018). Deep maps. In *This Is Not an Atlas*, Kollektiv Orangotango+ (ed.). Transcript Verlag, Bielefeld.

BnF Gallica (2020). La carte des Gallicanautes – uMap [Online]. Available at: https://umap.openstreetmap.fr/fr/map/la-carte-des-gallicanautes_409214#6/.

Bodenhamer, D.J. (2018). Chasing Bakhtin's ghost: From historical GIS to deep mapping. In *The Routledge Companion to Spatial History,* Gregory, I., DeBats, D., Lafreniere, D. (eds). Routledge, London, New York.

Bodenhamer, D.J. (2021). The varieties of deep maps. In *Making Deep Maps: Foundations, Approaches, and Methods,* Bodenhamer, D.J., Corrigan, J., Harris, T.M. (eds). Routledge, London, New York.

Bodenhamer, D.J., Corrigan, J., Harris, T.M. (eds) (2010). *The Spatial Humanities: GIS and the Future of Humanities Scholarship*. Indiana University Press, Bloomington and Indianapolis.

Bodenhamer, D.J., Corrigan, J., Harris, T.M. (eds) (2015). *Deep Maps and Spatial Narratives, Bloomington & Indianapolis*. Indiana University Press, Bloomington and Indianapolis.

Brando, C., Frontini, F., Moreau, D., Roche, M., Masson, É. (2021). Introduction. Humanités numériques et analyses spatiales : enjeux et perspectives. *Humanités Numériques*, 2021, 3.

Bunel, M., Olteanu-Raimond, A.-M., Duchêne, C. (2019). Objets et relations spatiales composites et prise en compte du vague pour interpréter un référencement spatial indirect. *Revue Internationale de Géomatique*, 29, 81–106.

Conseil départemental de la Seine-Saint-Denis (2024). Atlas de l'architecture et du patrimoine [Online]. Available at: https://patrimoine.seinesaintdenis.fr/Le-patrimoine-en-Seine-Saint-Denis.

Cooper, D. and Gregory, I.N. (2011). Mapping the English Lake District: A literary GIS. *Transactions of the Institute of British Geographers,* 36, 89–108.

Cope, M. and Elwood, S. (eds) (2009). *Qualitative GIS: A Mixed Methods Approach*. SAGE Publications Ltd, London.

Corrigan, J. (2015). Genealogies of emplacement. In *Deep Maps and Spatial Narratives*. Bodenhamer, D.J., Corrigan, J., Harris, T.M. (eds). Indiana University Press, Bloomington and Indianapolis.

Cousin, G. (2023). Déporter à pas d'hommes. Le transfert des Roms itinérants en Transnistrie, 1942. *Revue d'Histoire de la Shoah*, 217, 397–418.

Debord, G. (1973). *La Société du spectacle*. Simar Films.

Debord, G. (2005). *Œuvres cinématographiques complètes*. Éditions Gallimard, Paris (3 DVDs).

Deflou-Leca, N. and Massoni, A. (2020). Col&Mon. Collégiales et Monastères de la France médiévale [Online]. Available at: https://colemon.huma-num.fr/.

Desbois, H. (2015). La carte et le territoire à l'ère numérique. *Socio. La nouvelle revue des sciences sociales,* 4, 39–60.

DH GIS Projects (2011). Anterotesis [Online]. Available at: https://anterotesis.com/wordpress/mapping-resources/dh-gis-projects/.

Di Lenardo, I. and Kaplan, F. (2023). Ce que les machines ont vu et que nous ne savons pas encore. *Sociétés et Représentations*, 55, 249–267.

Dominguès, C., Jolivet, L., Brando, C., Cargill, M. (2019). Place and sentiment-based life story analysis. *Revue Française des Sciences de l'Information et de la Communication,* 2019, 17.

Du Mouza, C. and Lamasse, S. (2022). Digital academic history and the challenge of uncertainty. In *XIth Heloise Workshop*, 12–13th May, Paris.

Elliott, T. (2018). The Pleiades gazetteer of ancient places: Inventing an "Un-GIS" and its uses. Lecture, École nationale des chartes, Paris.

Elwood, S. (2022). Toward a fourth generation critical GIS: Extraordinary politics. *ACME: An International Journal for Critical Geographies*, 21, 436–447.

Follin, J.-M., Simonetto, É., Chalais, A. (2021). Détection automatique des parcelles sur les plans napoléoniens : comparaison de deux méthodes. *Humanités Numériques,* 2021, 3.

Geomedialab (2018). Atlasciné [Online]. Available at: https://rs-atlascine.concordia.ca/index.html.

Gioux, M. (2018). Géolocaliser les documents numérisés avec Gallicarte [Online]. Available at: https://gallica.bnf.fr/blog/21032018/gallicarte-arrive-dans-gallica?mode=desktop.

Goguey, F.M.M. (2022). Opacité et transparence dans le design d'un dispositif de surveillance urbain : le cas de l'IMSI catcher. PhD Thesis, EPFL, Lausanne.

Goodchild, M.F. (2015). Two decades on: Critical GIScience since 1993. *The Canadian Geographer/Le Géographe canadien,* 59, 3–11.

Grange, A. (2002). Promenade inférentielle d'un lecteur cartographe. Chapitre 10 d'Ulysse de James Joyce. *Communication et langages,* 132, 105–122.

Gregory, I., Bushell S., Cooper, D., Bolton, Z. (n.d.). Mapping the republic of letters [Online]. Available at: http://republicofletters.stanford.edu/casestudies/index.html.

Hägerstrand, T. (1970). What about people in regional science? *Papers of the Regional Science Association*, 24, 7–21.

Jakobson, R. (1963). *Linguistique et poétique. Essais de linguistique générale.* Éditions de Minuit, Paris.

Joliveau, T. (2015). Connecter espaces réels et espaces imaginaires grâce aux techniques géonumériques. Monde Géonumérique [Online]. Available at: https://mondegeonumerique.wordpress.com/2015/01/25/connecter-espaces-reels-et-espaces-imaginaires-grace-aux-techniques-geonumeriques/.

Kelly, M. and Bosse, A. (2022). Pressing pause, "doing" feminist mapping. *ACME: An International Journal for Critical Geographies*, 21, 399–415.

Kilouchi, S., Homo, J., Darty, K. (2023) Navigae : un outil polyvalent pour valoriser et visualiser les données géographiques patrimoniales. *Mappemonde*, 2023, 135.

Kwan, M.-P. (2008). From oral histories to visual narratives: Re-presenting the post-September 11 experiences of the Muslim women in the USA. *Social and Cultural Geography,* 9, 653–669.

Lancaster University (n.d.). Mapping the lake district: A literary GIS [Online]. Available at: https://www.lancaster.ac.uk/mappingthelakes/index.htm.

Le Noc, M. (2021). Towards a geography of persecution: The case of the Arts-et-Métiers and the Enfants-Rouges quarters of the third arrondissement of Paris, 1940-1944. PhD Thesis, Texas State University.

Least Heat-Moon, W. (1991). *PrairyErth: A Deep Map*. Houghton Mifflin Company, Boston.

Literary Map of NYC (n.d.). [Online]. Available at: https://literarymap.nyc/map/.

LiteraturAtlas (n.d.). Mapping and analysing the geography of fiction with interactive tools [Online]. Available at: https://www.literaturatlas.eu/en/index.html.

Mahmoudi, D. and Shelton, T. (2022). Doing critical GIS. *ACME: An International Journal for Critical Geographies,* 21, 327–336.

McDonough, K., Moncla, L., van de Camp, M. (2019). Named entity recognition goes to old regime France: Geographic text analysis for early modern French corpora. *International Journal of Geographical Information Science,* 33, 2498–2522.

McElroy, E. (2022). Digital cartographies of displacement: Data as property and property as data. *ACME: An International Journal for Critical Geographies,* 21, 357–371.

Moncla, L., Gaio, M., Nogueras-Iso, J., Mustière, S. (2016). Reconstruction of itineraries from annotated text with an informed spanning tree algorithm. *International Journal of Geographical Information Science,* 30, 1137–1160.

Morel, J. (2019). Modélisation spatiale du récit littéraire complexe de Kateb Yacine. *Mappemonde*, 2019, 125.

Morel, J. and Hautdidier, B. (2023). Pour un dialogue entre la géomatique et les humanités numériques : propositions pour contribuer à des pratiques géo-numériques critiques. In *SAGEO 2023*, Laval, Quebec [Online]. Available at: https://crdig.ulaval.ca/wp-content/uploads/actes_conference_sageo_2023.pdf, 251-267.

Morel, J., Crouzevialle, R., Massoni, A. (2017). Atlas Historique du Limousin accueil [Online]. Available at: https://www.unilim.fr/atlas-historique-limousin/.

Morel, J., Crouzevialle, R., Massoni, A. (2020). Construction and management of a geo-historical information system for an interdisciplinary and contributory atlas: The historical atlas of Limousin. *International Journal of Humanities and Arts Computing*, 14, 27–45.

Moretti, F. (2000). *Atlas du roman européen (1800-1900).* Le Seuil, Paris.

Moretti, F. (2005). *Graphs, Maps, Trees: Abstract Models for Literary History*. Verso Books, London.

Moscati, P. (2021). La raccolta bibliografica di "archeologia e calcolatori": La formazione. In *La bibliografia di informatica archeologica nella cultura digitale degli anni novanta,* Caravale, A. and Moscati, P. (eds). All'Insegna del Giglio, Sesto Fiorentino.

Murrieta-Flores, P. and Martins, B. (2019). The geospatial humanities: Past, present and future. *International Journal of Geographical Information Science,* 33, 2424–2429.

Noizet, H., Bove, B., Costa, L. (eds) (2013). *Paris de parcelles en pixels. Analyse géomatique de l'espace parisien médiéval et moderne*. Presses Universitaires de Vincennes, Comité d'histoire de la ville de Paris, Paris.

Odie, C. (2017). *Vectorisation semi-automatique de planches scannées du cadastre ancien*. Mémoire d'ingénieur CNAM, ESGT, Le Mans.

Pavlovskaya, M. and St. Martin, K. (2007). Feminism and geographic information systems: From a missing object to a mapping subject: Feminism and geographic information systems. *Geography Compass*, 1, 583–606.

Perret, J., Gribaudi, M., Barthelemy, M. (2015). Roads and cities of 18th century France. *Scientific Data*, 2, 150048.

Peuquet, D. (1994). It's about time: A conceptual framework for the representation of temporal dynamics. Geographic information systems. *Annals of the Association of the American Geographers,* 3, 441–461.

Piatti, B., Reuschel, A.K., Bär, H.R., Hurni, L. (2008). Die geographie der fiktion – das projekt "Ein literarischer Atlas Europas". *Kartographische Nachrichten,* 6, 287–294.

Presner, T. and Shepard, D. (2015). Mapping the geospatial turn. In *A New Companion to Digital Humanities*, Schreibman, S., Siemens, R., Unsworth, J. (eds). John Wiley & Sons, Hoboken.

Reuscher, A.-K., Piatti, B., Hurni, L. (2012). Modelling uncertain geodata for the literary atlas of Europe. In *Understanding Different Geographies*. Kriz, K., Cartwright, W., Kinberger, M. (eds). Springer, Berlin, Heidelberg.

Rodier, X. and Saligny, L. (2010). Modélisation des objets historiques selon la fonction, l'espace et le temps pour l'étude des dynamiques urbaines dans la longue durée. *Cybergeo: European Journal of Geography, (Systèmes, Modélisation, Géostatistiques)*, 502 [Online]. Available at: http://journals.openedition.org/cybergeo/23175.

Talmey, A. (1969). Vladimir Nabokov talks about Nabokov. *Vogue*, December, 190–191.

Thuillas, O. (2010). GéoCulture : le Limousin vu par les artistes. *L'Observatoire,* 2(37), 82–85.

Time Machine (2024). About us [Online]. Available at: https://www.timemachine.eu/about-us/.

Travis, C. (2015). *Abstract Machine: Humanities GIS*. ESRI Press, Redlands.

Westerveld, L. and Knowles, A.K. (2021). Loosening the grid: Topology as the basis for a more inclusive GIS. *International Journal of GIS*, 35, 2108–2127.

Westphal, B. (2011). *La Géocritique: Réel, fiction, espace*. Minuit, Paris.

Wieck, L.P. (2018). Building inspector. *Journal of American History*, 105, 477–478.

Wood, D. (2010). *Everything Sings: Maps for a Narrative Atlas*. Siglio Press, South Egremont.

Wood, D. (2015). Mapping deeply. *Humanities*, 4, 304–318.

Wood, D. (2021). The art of deep mapping. In *Making Deep Maps: Foundations, Approaches, and Methods*, Bodenhamer, D.J., Corrigan, J., Harris, T.M. (eds). Routledge, London, New York.

Young, S.J. and Levin, J. (n.d.). Mapping St Petersburg [Online]. Available at: https://www.mappingpetersburg.org/site/.

Zwer, N. (ed.) (2023). *Ceci n'est pas un Atlas*. Éditions du Commun, Rennes.

7

Uses and Unconsidered Aspects of Geographic Data in OSINT Investigations

Allan DENEUVILLE
UR MICA, Université Bordeaux Montaigne, France

7.1. Introduction

Over the past few decades, the geodigitization of the world has gradually established itself as an essential pillar of contemporary digital investigations. These investigations rely on the ingenious exploitation of vast amounts of data generated by users on online platforms, ranging from social media to websites and publicly accessible geographic databases. Digital technology has provided a deluge of open data, enabling amateur and professional investigators alike to not only explore topics such as the territorial advance of armed conflicts, but also to document the ecocidal actions of states and companies, and to expose police violence and the illegal deportation of asylum seekers. This open-source research methodology is known as OSINT, or Open Source INTelligence. It encompasses investigations based on all freely accessible online data. However, there is some debate about the limits of this approach, for example in the case of information leaks or the purchase of documents on the black market. This has been illustrated by an investigation by Bellignat, which led to the identification of FSB agents (Russian intelligence services)

Geospatial Technologies and Society,
coordinated by Thierry JOLIVEAU and Matthieu NOUCHER. © ISTE Ltd 2026.

involved in the attempted poisoning of Russian political opponent Alexei Navalny in 2020[1].

OSINT is part of the broader landscape of intelligence techniques, alongside MASINT (Measurement And Signature INTelligence), SIGINT (SIGnals INTelligence), HUMINT (HUMan INTelligence), and other forms of GEOINT (GEOspatial INTelligence). These approaches offer a complex and multifaceted overview of contemporary reality. Although OSINT has its historical roots in the methods of espionage and intelligence analysis prevalent before and during World War II, recent research suggests that this practice may date back even further, to the early 20th century in Europe and even to the 19th century in the United States (Block 2023). Nevertheless, it was not until the late 20th and early 21st centuries that these techniques experienced considerable growth and democratization, with the emergence of social media on the one hand and the accessibility of digital maps and satellite imagery on the other. This was due to the deployment of private companies offering these services, free or fee-based. In this context, many stakeholders, including news media, non-governmental organizations and human rights associations, have developed online guides to train investigators and citizens in the collection and analysis of geospatial data (*Agence France-Presse*[2], *Al Jazeera*[3], *Global Investigative Journalism Network*[4], *Tactical Tech*[5], etc.). The most important of these is undoubtedly the Berkeley Protocol (United Nations and University of California, Berkeley 2022). These resources play an essential role in broadening access to information and empowering individuals, developing their ability to understand and act on the complex issues of our time. These individuals can then become stakeholders in investigations concerning their own situations. As *Al Jazeera* writes in its guide to using satellite imagery in open source investigations: "Satellite imagery has also become a compelling centerpiece for visual storytelling, and a window into remote or restricted locations. Investigative journalists can use

1 Bellingcat Investigation Team (2020). Hunting the hunters: How we identified Navalny's FSB stalkers. *Bellingcat*, 14 December [Online]. Available at: https://www.bellingcat.com/resources/2020/12/14/navalny-fsb-methodology/.

2 Agence France Press (n.d.). Geolocation 1. *Agence France Press* [Online]. Available at: https://fr.digitalcourses.afp.com/courses/identifiez-les-indices-d-une-image-voyagez-dans-le-temps.

3 Creta, S. (2023). How to analyse satellite imagery. Al Jazeera Media Institute, 25 May [Online]. Available at: https://institute.aljazeera.net/en/ajr/article/2191.

4 McNeill, R., Nelson, D., McIntosh, T. (2018). Resources for finding and using satellite images. *Global Investigative Journalism Network*, 5 December [Online]. Available at: https://gijn.org/fr/ressource/guide-pour-trouver-et-utiliser-des-images-satellites/.

5 Gutermuth, L. (2016). Starting satellite investigations. Exposing the invisible, 27 September [Online]. Available at: https://exposingtheinvisible.org/en/guides/starting-satellite-investigations/.

satellite imagery to make visible what governments or institutions want hidden out of sight." Thanks to OSINT, many investigators aim to act as a counter-surveillance watchdog on state power and to conduct counter-investigations (Deneuville et al. 2022) using satellite images. Investigators have taken advantage of the opening up of the market for satellite imaging devices and the end of the exclusivity of cartographic data previously held by states and the military establishment. This has enabled them to monitor the power that had previously shaped reality through its representations of the world (Boltanski 2011). This surveillance was accompanied, albeit with a delay of several decades, by a reversal of perspective brought about by the broader uptake of photographic and then videographic recording devices, and the ability to disseminate them globally, made possible by the emergence of social media. Admittedly schematic – since mapping has never been exclusively in the hands of states or the military, for example – this sharing makes it possible to account for the way in which many OSINT investigators see themselves.

Although ubiquitous in OSINT investigations, the production and use of images from mapping software carries with it a set of conceptual and political blind spots. These are undoubtedly inherent in any use of images presented as authoritative evidence, but whose production processes are not deconstructed and analyzed. To illustrate these blind spots, in this chapter we will focus on two types of OSINT investigations: investigative journalism and human rights and environmental advocacy reports. We will draw on examples of specific investigations to study what is at stake in their circulation. Here, we will discuss digital maps that translate the avalanche of digitally produced data to account for the territory. By metonymy, we may be tempted to mention only Google Maps and Google Earth. This may seem relevant, since these digital applications have already been analyzed by researchers to reveal the semiotic and political intricacies behind their user-friendly interface (Joliveau 2016; Chasseray-Peraldi 2022). However, as we shall see, OSINT investigations also draw on other resources that complicate the political issues surrounding the use of these web maps (Noucher 2017) and this cartography 2.0 (Joliveau et al. 2013). The case of Google remains interesting in that it crystallizes the tensions inherent in this type of cartographic use. This central player in Big Tech, which became Alphabet in 2015, can be seen as a tool for the potential emancipation of citizens, since it gives them access to information. Thus, in an article on migration routes and crossing the Darién Gap, the authors quote a migrant who confidently states that "Google Maps is my smuggler" (Sarrut et al. 2023). However, this may be jumping to conclusions, if we recall the history of this application's deployment and its harmful logic, as explored by Shoshana Zuboff in her book *The Age of Surveillance Capitalism* (2018). This tension echoes the warning of African-American poet Audre Lorde: "the master's tools will never dismantle the master's house" (2018). Indeed, even if they sometimes support temporary struggles, these master tools encode a vision imposed by those in power in the actions they propose. It is for this reason that other bottom-up mapping systems, set up by citizens, are developing in response to the top-down power imposed by a ubiquitous multinational corporation. This is also evidenced by the development of radical citizen mapping technics (Zwer and Rakacewicz 2021). When using this type of map, OSINT investigators are not immune to the regime of truth and governmentality (Foucault 2007, 2016), as these applications function as algorithmic black boxes (Noucher 2023), filtering information and thus acting as gatekeepers (Barzilai-Nahon 2008) who decide what is made visible and what is made invisible.

To address these issues, the first part of this chapter will draw on journalistic investigations and NGO reports to examine the different ways of investigating with geographic data and using it in open source investigations. We will then look at the different regimes of truth in which these maps are embedded. Finally, we will study how, based on the specific case of an investigation into borders, the perspective of power encoded in online maps can lead investigators to conceptual and political blind spots.

7.2. Investigations and maps

Although OSINT has been around for many years, as we mentioned, it has taken a decisive turn with the enormous amount of data made available by Web 2.0 and the exponential growth of social media, whose audiovisual content publication figures are simply staggering: every minute, 500 h of video are posted on YouTube, 65,000 photos are posted on Instagram and 240,000 on Facebook, to name just a few of the most popular networks in the West[6]. We can broadly outline three ways of using digital maps in open source investigations: the map as a tool, as an investigation's starting point and as an illustration. Of course, this division is only for didactic and demonstrative purposes, as these three registers are intertwined throughout the investigations.

7.2.1. *The map as a tool for investigation*

One of the first OSINT investigations to receive widespread media coverage and rely heavily on digital mapping is *Anatomy of a Killing*, produced by BBC Africa Eye in September 2018. It is an investigation into the summary execution of two

6 See: https://web-assets.domo.com/miyagi/images/product/product-feature-22-data-never-sleeps-10.png.

women and two children in a desert area, allegedly in Mali. Using a video lasting around two minutes, journalists Aliaume Leroy and Ben Strick, accompanied by their teams, managed to geolocate the video in Cameroon and date it with a margin of error of around two weeks. To do this, they started with a clue from a ridge line in the background of the video and managed to match it with their research on digital maps. Based on the presence or absence of elements in the video, compared with satellite images, and using calculations of shadows cast on the ground, they were then able to date it. This investigation highlights one of the key uses of digital maps and satellite images in OSINT investigations: fact-checking (Bigot 2019). This type of investigation is also frequently conducted during armed conflicts, with the aim of countering propaganda in information wars, as we have recently seen with the Russian Federation's invasion of Ukraine. *The New York Times*, for example, produced a major open-source investigation into the Bucha massacre (Deneuville 2024a). Geospatial data in OSINT investigations needs to be cross-referenced with temporal data also provided by online maps. Indeed, a significant portion of investigations are based on a before/after comparison to demonstrate the consequences of an event by pointing out the traces it has left on the territory. This visual form was already in use in the early 2000s, as Laura Kurgan recalls in her study of the first satellite images published by *The New York Times* in 2000. The newspaper used images from the Ikonos satellite to document the destruction of the Chechen capital Grozny during the Chechen War and at the dawn of Vladimir Putin's election (Kurgan 2013). It juxtaposed images taken four months apart.

Used in this way, the map makes it possible to investigate human rights crimes without endangering journalists or investigators. The latter can also explore digital territories from which they are excluded for political reasons. This is the case, for example, for Western academics currently working on Russia, who are banned from entering the territory of the Federation, as Audinet and Limonier (2022) show in their issue of the journal *Hérodote* on digital terrain. Beyond OSINT, investigations can also begin with online research as preparatory groundwork to assess the relevance of sending investigators into the field. Depending on the speed and position of satellites, newsrooms can obtain satellite coverage of an area where an event they may or may not wish to cover has just occurred within a matter of hours.

7.2.2. *The map as the basis for the investigation*

Other investigations use maps not to verify information, but rather as a starting point. In this type of approach, the aim is often to find information on maps that are accessible to everyone or, conversely, on maps whose existence and wealth of information are unknown to those who do not know how to use them. One example is the work carried out by *The New York Times*[7] and *Médiapart*[8], which, through the ingenious use of the heat map from the Strava running app, identified the location of unofficial military bases in Africa and Asia belonging to the US and French armies. The Strava app has a heat map[9] that shows the joggers' running routes around the world. Although the app is mainly used in the West, there are some colored areas on the African continent and in parts of Asia, and if you zoom in on the map, you can see areas in the heart of the Sahel and in Afghanistan. By cross-referencing this information with satellite images, journalists have found that these areas correspond to the outlines of military bases, some of which are secret.

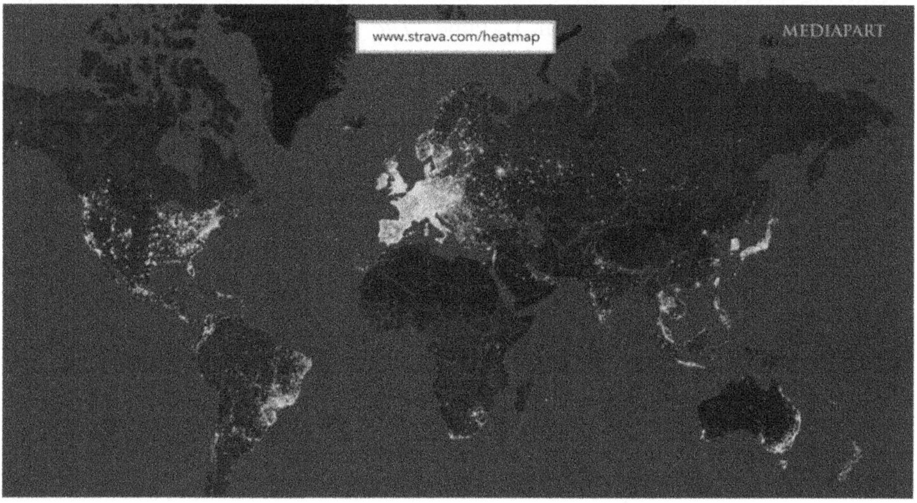

Figure 7.1. *Screenshot from the investigation "French soldiers compromise the security of their operations on social media", Médiapart, December 10, 2020. Heat map from the Strava running app. For a color version of this figure, see www.iste.co.uk/joliveau/geovisualization.zip*

7 New York Times (2018). How Strava's heat map uncovers military bases. *New York Times*, 30 January [Online]. Available at: https://www.nytimes.com/video/world/middleeast/100000005705502/big-data-big-problems-how-stravas-heat-map-uncovers-military-bases.html.

8 Brabant, J., Bourdon, S. (2020). French soldiers are compromising the security of their operations on social media. *Médiapart*, 10 December [Online]. Available at: https://www.mediapart.fr/journal/international/101220/des-militaires-francais-compromettent-la-securite-de-leurs-operations-sur-les-reseaux-sociaux.

9 See: https://www.strava.com/maps/global-heatmap.

From there, journalists can then identify soldiers, their relatives and their home addresses. This reminds us how the surveillance system put in place by ubiquitous digital technology can potentially turn against everyone. Indeed, if journalists can uncover this information from their computers, it is highly likely that the geopolitical enemies of the countries in question had already found it themselves. These investigations invite us to question the need today to keep certain areas of maps blank (Noucher 2023), or the political and epistemological issues behind the preservation of digital traces cross-referenced with geographic data, or even the risk of reducing the world to the traces we capture in the illusion of reality they give us (Beaude 2015). Nevertheless, these investigations also demonstrate how a journalistic investigation can be based on data found on an online map.

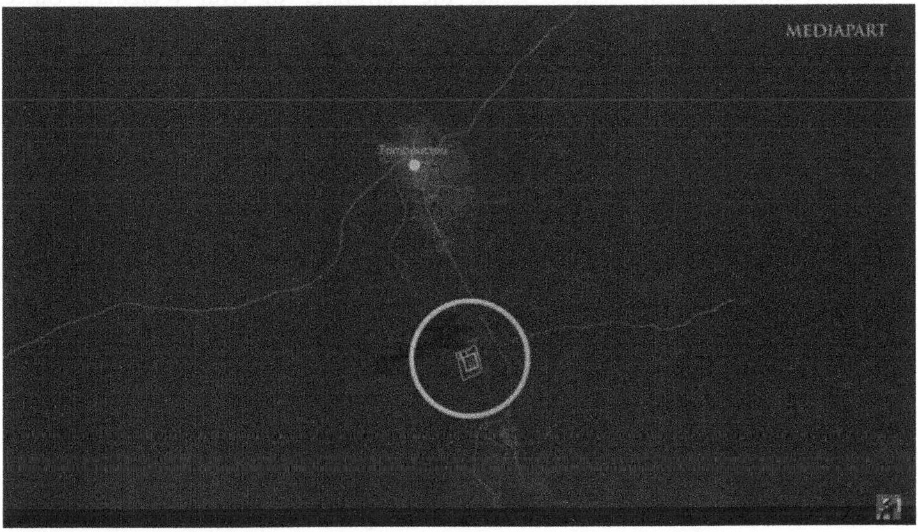

Figure 7.2. *Screenshot from the investigation "French soldiers compromise the security of their operations on social media", Médiapart, December 10, 2020. The enlarged map shows the outline of an unofficial French army base*

Other investigations based on this model of data interpretation may also be conducted: investigations into ecocide and climate change are prime examples. A number of satellites take images across multiple spectral wavelengths, providing access to a set of data invisible to the naked eye over a fairly long period of time. The quality of the information provided by a satellite image therefore depends on its multispectral capacity. Satellites are equipped with sensors that record light from multiple distinct wavelengths, covering not only the visible spectrum but also infrared. This type of information is publicly available, for example, on maps and satellite images on the NASA website[10]. These investigations can operate on the demonstrative principle of before/after, which we have already discussed and which is particularly effective in addressing issues such as deforestation, glacier melt and coral reef retreat. However, wavelengths also represent an important source of investigation into chemical pollution of the air, land and water, and its consequences on populations, by capturing traces of such pollution. One example is the investigations conducted by the non-profit organization Sky Truth into various environmental violations using satellite monitoring (offshore drilling, oil spills, hydraulic fracturing, mining, illegal fishing, etc.). We can also mention the impressive investigation published in 2019 by the Forensic Architecture collective. This is one of the leading figures in OSINT investigation, founded at Goldsmith–University of London under the patronage of Eyal Weizman. Since 2014, the collective has been working on land and field pollution in Gaza, documenting the spraying of herbicides by the Israeli army on the border with the Palestinian territories[11]. To assist with this type of investigation and to understand what can be expected from different satellite image providers, the Tactical Tech collective, through its "Exposing the invisible" project, has reproduced a fairly effective table, taken from a report by the NGO Poverty Action, of the different operators and the different wavelengths available that they offer[12].

7.2.3. The map as an illustration of the investigation

In addition to these two ways of investigating using online maps, we also note their omnipresence in OSINT investigation videos. They have become one of the aesthetic patterns that make this methodology recognizable and are an integral part of a "visual grammar of veridiction" (Deneuville 2024b). We thus establish a conceptual distinction between Foucauldian "regimes of truth" (Foucault 2011, p. 27), which represent the mechanisms for producing and manifesting truth in a society, and "veridiction", which does not necessarily produce a discourse of truth but nevertheless claims to deliver one. While this use of maps often allows viewers to follow the investigator step by step in their investigation, they can also serve as simple illustrative rhetorical devices. Maps can indeed contribute to the storytelling of the investigations carried out, with the function of capturing the viewer's

10 See: https://worldview.earthdata.nasa.gov/.

11 Forensic Architecture (2019). Herbicidal warfare in Gaza. *Forensic Architecture*, 17 July [Online]. Available at: https://forensic-architecture.org/investigation/herbicidal-warfare-in-gaza.

12 Gutermuth, L. (2016). Op. Cit.

attention and persuading them. As Micah Farfour says in the reference book for OSINT investigators, *Digital Witness*:

> News and other media use satellite imagery to demonstrate analysis and provide visuals to tell the story. While words can be engaging, satellite imagery can help to explain the story, allowing to visualize and potentially relate better to the story. There is also possible information within the data from a satellite image that can explain the situation better than any words. (Dubberley et al. 2020, p. 232).

This is an aspect that Google Earth understands well, allowing its users, on its Google Earth Pro and Google Earth Studio versions, to create illustrations directly from the satellite images on its service, so that they can be used, for example, in OSINT investigation reports. However, the semiotic complexity of these illustrative maps and satellite images never seems to be addressed as it should be, reinforcing the impression of a conceptual blind spot in their use. One example of this is the use of satellite images in *The New York Times*' investigation about the Bucha massacre, sometimes simply as a backdrop to other elements. The resulting composite images then run the risk of losing viewers who may want to focus on their own details (Figure 7.3).

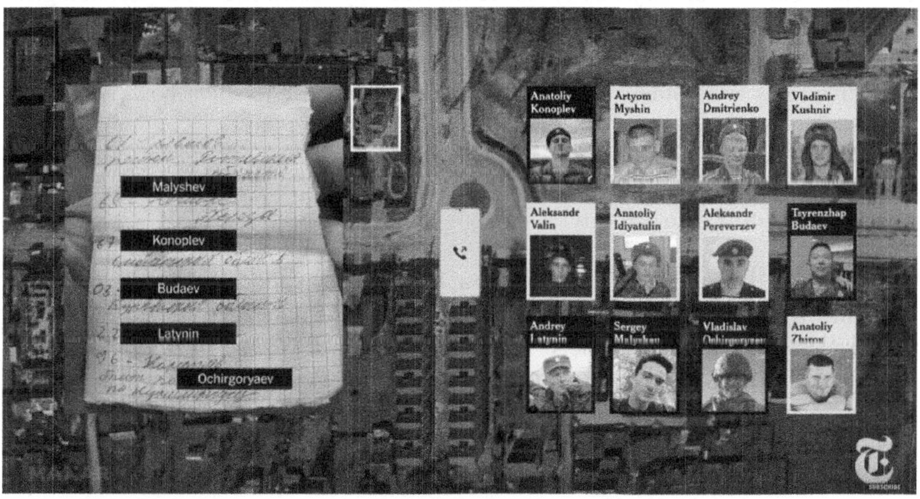

Figure 7.3. *Capture from the investigation Caught on Camera, Traced by Phone: The Russian Military Unit That Killed Dozens in Bucha, The New York Times, December 22, 2022. For a color version of this figure, see www.iste.co.uk/joliveau/geovisualization.zip*

Thus, we can see that digital maps and satellite imagery in OSINT investigations cover a range of different uses. Whether as tools, the basis for an investigation, or simple illustrations, the maps used by OSINT are demonstrative, probative and aesthetic all at once.

7.3. Truth regimes and maps

The maps used in investigations are rarely viewed as produced maps, as re-presentations of reality rather than simple, unmediated presentations of it. In fact, like any image-based production and like any data, these maps are what Latour (2007) proposes to call "obtained". Digital maps and satellite images offer social constructs rather than direct access to the phenomenological world. It therefore seems essential to question their impact on OSINT investigations that use these tools without taking them into consideration as such. First, it is useful to consider the regimes of truth within which these maps are situated. The aim is not to observe what could be hastily reduced to the conceptual blind spots of journalists and investigators, but rather to reveal the codes with which they are led to play in their investigations. Indeed, behind this use of maps in investigations lies a certain form of socially constructed culture of objectivity, as Daston and Gallison (2007) remind us. This culture of objectivity is reinforced by scientific and technological syncretism, which brings together cartographic science and photographic operators through digital maps. Both of these fields have a history of considering maps to be probative tools and holders of truth. We therefore propose to study how these maps fit into the regime of journalistic, legal and aesthetic truth.

7.3.1. *Truth and journalism*

It is indeed essential to consider how these investigations fit into the field of journalism. OSINT is expanding in a context where the dangers of fake news (Bigot 2019) and other "alternative facts" are being questioned (Esquerre 2018), as well as in a context of growing public mistrust of traditional media, as shown in Reuters reports (Newman et al. 2023). Some academics therefore consider OSINT to be a methodology whose investigative findings allow viewers to experience the different stages of the investigation step by step (Le Deuff and Roumanos 2021). They also note that OSINT, by tackling politically sensitive topics such as police violence and ecocide, could restore journalism's reputation as the fourth estate. This approach is accompanied more broadly by the media coverage of whistleblowers. Nevertheless, although the notion of journalistic truth is central to the contemporary media landscape, it is also often swept under the rug. Journalists are traditionally required to report facts accurately, impartially, and in a balanced manner, with the aim of providing a faithful representation of reality. This is what distinguishes the delivery of information from that of opinion. However, the truth referred to here cannot be considered unambiguous. As a media construct, journalistic truth is not completely out of sync with events occurring in the world. Its complexity defies any notion of absolutes. This truth, which is above all factual, stems from the authentication of facts and the establishment of a system of evidence specific to journalism, which is sometimes incompatible with other systems of evidence administration, such as the courts. Recently, we have seen OSINT investigations using geolocation being dismissed in criminal proceedings, such as when the Dutch public prosecutor's office suspended its investigation in 2023 into the crash of MH17 in 2014, which is one of the most well-known investigations by the OSINT investigative media outlet Bellingcat[13]. Nevertheless, the use of digital maps in OSINT investigations is part of a certain way of seeing the world and conceiving of journalistic responsibility with regard to factuality. Digital mapping allows causal links to be established through clusters of clues, in the Peircean sense of the term – that is, a sign that establishes a direct physical or causal connection with its object, left on the ground by the facts.

7.3.2. *Legal truth*

Journalism is not the only truth regime in which OSINT investigations and digital maps have a role to play. The use of maps also responds to certain logics and injunctions issued by legal authorities. Many OSINT investigations originate in newsrooms but may be taken into account by courts. Others are carried out directly by NGOs working with courts of law. More and more OSINT investigators are also being called upon as "experts" by the latter. This link with the legal sphere is reflected in the other name given to OSINT, coined by the Forensic Architecture collective: forensics, which refers to the legal sciences. These constitute a set of sciences used in the context of judicial investigations, the most famous of which is forensic medicine. There are many others, such as forensic linguistics, forensic psychiatry and computer forensics. Weizman (2017) defines forensics by broadening its scope and viewing it as the encounter between an object (a video, photo, satellite image, etc.) and an expert (an investigator, a physician, a geographer, etc.) who brings it to life before a forum (a court of law, a media, an art center, etc.). Thus, although we may criticize certain unconsidered aspects of the use of maps in OSINT investigations, we would like to mention at this point that some of these unconsidered aspects stem from OSINT's need to gain admissibility in the legal

13 Maupas, S. (2023). Crash of flight MH17: International investigation team suspends its inquiry. *Le Monde*, 8 February [Online]. Available at: https://www.lemonde.fr/international/article/2023/02/08/crash-du-vol-mh17-les-enqueteurs-internationaux-suspendent-leur-enquete_6161074_3210.html.

truth regime. They would then reflect a compromise reached by investigators, resulting from their desire to lend credibility to the OSINT methodology in the eyes of the judicial systems, by complying with certain ways of administering evidence. In this tense situation, as the Human Rights Center at the University of California, Berkeley points out, there is a struggle to have OSINT investigations accepted as sufficient evidence (United Nations and University of California, Berkeley 2022). Satellite imaging technologies can play a key role here, for example, through the resolution of the images presented to the court. Citizens do not have access to images with the same high definition as those used by the military. The resolution of images for commercial equipment in conventional channels is 31 cm per pixel, but not necessarily in all areas and at all times. Nevertheless, for open source evidence from investigations to be accepted, a certain image resolution is often required for it to be accepted by the courts (Dubberley et al. 2020, p. 228). Indeed, physical "traces" on a territory, if below the resolution limit offered by satellite images, for example, if they are smaller than a pixel, are not visible and therefore cannot be admissible evidence in legal proceedings. This is why Eyal Weizman proposes talking about a "detectability threshold" that open source investigators must negotiate to report on the facts they want to document using satellite images. For example, a state may attempt to limit the resolution of this type of image to conceal certain traces left on its territory. While this may be understandable for military security reasons, such as making operational bases invisible, it is more questionable when US imaging satellites have an image resolution for Israeli and Palestinian territories that is several times lower than that for other states, effectively making a set of physical traces left on these highly conflictual territories invisible (Weizman 2017).

7.3.3. *Visual grammar of veridiction*

In the face of these two types of discourse, OSINT investigations give rise to a form of surreptitious veridiction that is conveyed through the visual forms used to report on the investigations, which we have proposed to call the "visual grammar of veridiction" (Deneuville 2024b). Indeed, through the video reconstruction of these investigations, we have the impression that the journey we take with these investigations leads us from point A to point B as if by fate, transforming each visual scheme used into a tool in the service of scientific demonstration. We have already discussed the use of before/after images in online maps, which, through the juxtaposition of images, seem to prove an event in a semiotic interweaving that is more complex than it appears. We can also note the recurring use of the juxtaposition of satellite images with images, videos or photos taken from the ground. This combination creates a spatial projection that reinforces the truthfulness of the map. Frederik Bjerknes writes about investigations carried out by *The New York Times*:

> The maps are perceived as real, not because of their correspondence to reality, but due to their correspondence with the adjacent video footage. Simultaneously, the truth-value of the footage is strengthened vice versa by the very same logic: When NYT pinpoints on maps where the footage was taken, coherence is made between the two artifacts, generating a dynamic where they validate each other. (Bjerknes 2022)

To strengthen the links between the image taken from the ground and the satellite image, geometric figures of different colors are used to create correspondences that appear obvious. This is what we see, for example, in Figure 7.4, taken from a 2022 BBC investigation entitled *Death on the Border*[14], which focuses on the deaths of dozens of exiles at the Moroccan-European border in Melilla, to which we will return later.

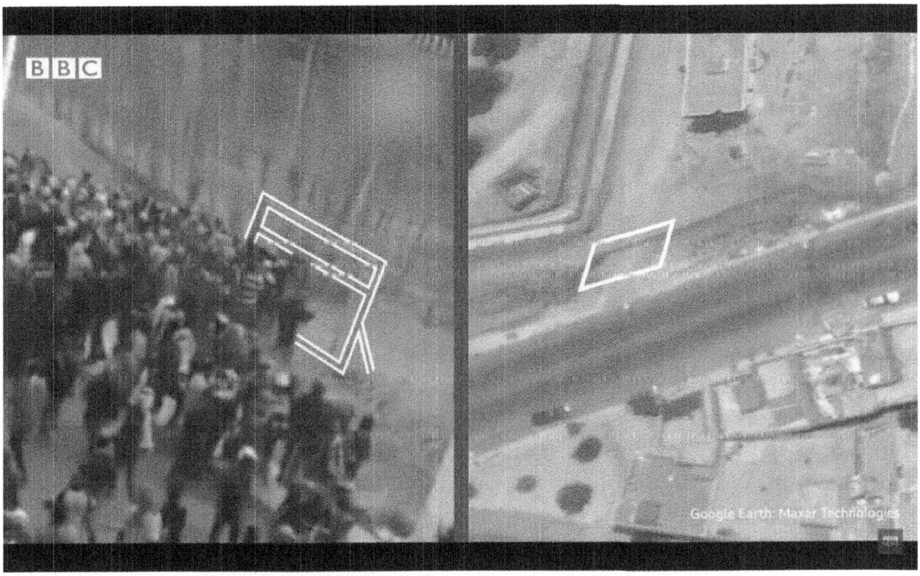

Figure 7.4. *Screenshot from the BBC investigation Death on the Border, November 6, 2022*

14 BBC Africa Eye (2022). Death on the border. *BBC News*, 6 November [Online]. Available at: https://www.bbc.co.uk/programmes/m001f2h2.

7.4. The risk of technological and political determinism

Within these regimes of truth, we may be tempted to think that the map alone is sufficient to tell the truth, provided we know how to extract the right information from it and that there is no possibility of manipulation. However, while inexperienced investigators may be fooled by this illusion, the more advanced among them are well aware of the limitations of digital map and satellite image research. Nevertheless, there are still some blind spots and unconsidered aspects of these uses that we feel are important to point out. To do so, we will examine how errors can find their way into maps, before focusing on the ways in which maps can steer investigations toward a certain technological determinism that carries with it an ideological bias that must be taken into account.

7.4.1. *Misinterpreting the map*

A map never says anything on its own. It is how it is interpreted that determines whether or not it is a tool for seeking the truth. As Micah Farfour reminds us: "Satellite imagery, in most situations, cannot independently prove an abuse has been committed, but it can add – sometimes essential – information to what may have transpired" (Dubberley et al. 2020, p. 233). To consider the type of errors found in the maps, or rather in their interpretations, we can refer to Fitterelli's article in Bellingcat reviewing an investigation into Chad[15]. The author describes how he and his colleagues could have been misled by reading the maps too hastily, wanting to see pre-established scenarios. By cross-referencing their analyses with other sources, they realized they had almost fallen into confirmation bias. Upon investigating reports that the Chadian National Army was setting fire to areas of the islands in Lake Chad to drive out opposing fighters, they initially observed that these reports were consistent with satellite images and videos. However, upon further research, they learned that these fires are a regular occurrence on these islands. Triangulation of information is therefore essential. Cross-referencing sources is necessary to avoid falling into the trap of spurious correlations that arise when images and maps are superimposed. We wrote earlier that maps and images were the legacy of a tradition that viewed them as irrefutable evidence. However, many image theorists have long analyzed and criticized this position (Lavoie 2017; Ledoux and Zabunyan 2023), as have many geographers (Harley 1989; Pickles 1995). While this research often seems unknown to OSINT investigators, who are sometimes blinded by the

15 Fittarelli, A. (2020). Ghost in the machine: From Chad, a case study on why you shouldn't blindly trust tech. *Bellingcat*, 17 July [Online]. Available at: https://www.bellingcat.com/resources/2020/07/17/ghost-in-the-machine-from-chad-a-case-study-on-why-you-shouldnt-blindly-trust-tech/.

persuasive power of these two types of evidence, it is important and urgent to take note of it at a time when images generated by artificial intelligence can be used to create "location spoofing" (Zhao and Sui 2017; Zhao et al. 2021) and lead OSINT investigators down misleading paths.

7.4.2. *Technological determinism*

Alongside these errors in interpreting maps, there is a risk of overlooking the way in which maps filter information and impose a certain worldview. McLuhan (1964) writes that technologies shape us. German philosopher Kittler (1999) goes further by proposing the idea of radical technodeterminism. In the opening of his book *Gramophone, Film, Typewriter*, he writes: "The media determine our situation. And yet, or precisely for this reason, this situation must be described". While relativizing these positions, whose main flaw is to underestimate the social and political factors in which technologies are embedded, we can nevertheless identify some risks of technological determinism hanging over investigations using digital maps. To do so, we can start with the BBC's *Death on the Border* investigation, which we have already discussed. This video recounts the tragic events that took place on June 24, 2022, involving nearly 2,000 refugees who attempted to cross the border between Morocco and the Spanish enclave of Melilla. Trapped against the border walls, they were brutally pushed back by Moroccan law enforcement, resulting in the loss of human lives. At least 24 people died, while many others remain missing. Some refugees found themselves trapped in an inner courtyard at the border, where they were attacked while lying on the ground and posing no threat. Others finally reached European soil but were illegally turned back by Spanish authorities before they could apply for asylum. The investigation, supported by witness statements, focused at length on the question of whether the exiles had died on European or Moroccan soil, analyzing the position of their bodies in relation to the border. The aim was to determine the responsibility of each country, no doubt with a view to providing legal assistance. However, when the video investigation is viewed, it creates a certain unease due to the cold gaze cast upon these bodies. How close to the border must a body be located in order for a state to be held responsible? Does it make any difference if a person dies 10 cm further north or south? Latour (2007) explained that scientific tools, as well as scientific language, construct the reality of the field. Thus, in the world of Google Earth, borders are represented as tangible realities, clear lines on the ground that are not pixelated, unlike satellite images. The problem with OSINT, when presented as geographical truth, lies in the lack of critical distance from the object to be deconstructed. Maps, even digital and satellite maps, are subjective discourses with pre-established intentions, whether conscious or unconscious. They encode a certain worldview and political choices.

Maps do not represent the territory itself (Korzybski 1995). Thus, the borders proposed by Google Earth representations change depending on the country from which the application is consulted. By focusing on the geography of the border from digital maps, OSINT leaves it unquestioned. This use of maps somehow reenacts a form of technological positivism that is worth distancing ourselves from in light of the current migration crises.

7.4.3. *A vision of power encoded in the map*

More than technological determinism, these maps encapsulate a certain unquestioned political vision. By implication, they naturalize borders and hinder the perception of them as arbitrary social constructs, while at the same time helping to reinforce a collective imagination around them, even in critical discourse. As Amilhat-Szary points out: “the proliferation of discourse, even critical discourse, on borders contributes to the media’s reinforcement of the idea that borders exist permanently” (Amilhat-Szary 2015, p. 101). Furthermore, this type of cartographic representation distances us from the event and seems to question the relevance of being on the ground. By mediating it, it offers an approximation of the event (Parks 2018). Indeed, while it is never nighttime on Google Maps, we never see social movements expressed either. It cannot transcribe the reality of the social world in all its complexity. These maps thus refer to a certain bird’s-eye view, which we could call drone-like (Chamayou 2015). This view carries with it a certain vision of the world and of the place of subjectivities and social agents within events. In line with data journalism, the use of digital maps and satellite imagery, which render the links between citizens invisible, may run the risk of coldly dehumanizing the field. These OSINT investigations also show us how these satellite maps are also tools of war, echoing Lacoste’s (1976) famous quote and book title: “Geography is used, first and foremost, to wage war”. For OSINT investigations, this can involve information warfare, but also armed conflict, allowing for the precise mapping of the designated enemy. Investigations based on images from the Maxar satellite showed that the word “child” in Russian had been written on the ground around the theater in Mariupol that was bombed by the Russian Federation’s armed forces. This is not to say that journalists and other OSINT investigators do the same thing as the military does, but to keep in mind that the same tools are used and have been built, and therefore retain within themselves a certain political view of the world. We are therefore invited to think not about how to use them for purposes other than those for which they were built, but to find ways to subvert their processes. Heller and Pezzani remind us of this in their important text on an investigation into the left-to-die boat:

> In a context where remote sensing is so central to the process of illegal immigration and where the success of a clandestine border crossing depends in part on not being detected, how can we avoid becoming accomplices to the government's attempt to control migration by highlighting border transgressions? (Heller and Pezzani 2014)

Thus, we can ask ourselves, along with OSINT investigators, how to use these tools without being complicit in the harmful logic encoded within these algorithmic black boxes.

7.5. Conclusion

In conclusion, having noted the diversity of uses for digital maps and satellite images in open source investigations, we have considered how, while falling within a set of different truth regimes, the use of these tools may be subject to a number of conceptual and political blind spots in the results of these investigations. However, this does not necessarily mean that the use of satellite maps in OSINT investigations should be abandoned. In an analysis inspired by Plato, the French philosopher Derrida (1993) compares writing to a pharmakon, a substance that is both a remedy and a poison depending on the dosage. Perhaps the use of maps falls within a much broader and more complex political spectrum than radical technological or political determinism.

7.6. References

Amilhat-Szary, A-L. (2015). *Qu'est-ce qu'une frontière aujourd'hui ?* PUF, Paris.

Audinet, M. and Limonier, K. (eds) (2022). OSINT : enquêtes et terrains numériques. *Hérodote*, 186, 192.

Barzilai-Nahon, K. (2008). Toward a theory of network gatekeeping: A framework for exploring information control. *Journal of the American Society for Information Science and Technology*, 59(9), 1493–1512.

Beaude, B. (2015). Spatialités algorithmiques. In *Traces numériques et territoires*, Severo, M. and Romele, A. (eds). Presses des Mines, Paris.

Bigot, L. (2019). *Fact Checking vs Fake News. Vérifier pour mieux informer*. Éditions de l'Institut National de l'Audiovisuel (INA), Paris.

Bjerknes, F. (2022). Images of transgressions: Visuals as reconstructed evidence in digital investigative journalism. *Journalism Studies*, 23(8), 951–973.

Block, L. (2023). The long history of OSINT. *Journal of Intelligence History*, 23, 1–15.

Boltanski, L. (2011). *On Critique: A Sociology of Emancipation*. Polity, London.

Chamayou, G. (2015). *Drone Theory*. Penguin, London.

Chasseray-Peraldi. P. (2022). Les paysages numérisés de Google. De l'environnement des images aux coexistences perceptives. PhD Thesis, CELSA-Université Paris Sorbonne, Paris.

Daston, L. and Galison, P. (2007). *Objectivity*. Zone Books, Princeton.

Deneuville, A. (2024a). Revenir sur Boutcha – Désinformer la guerre en Ukraine avec l'OSINT. Analyse Opinion Critique (AOC) [Online]. Available at: https://aoc.media/analyse/2024/05/12/revenir-sur-boutcha-desinformer-la-guerre-en-ukraine-avec-losint/.

Deneuville, A. (2024b). (Contre-)enquêtes OSINT, Publictionnaire. Dictionnaire encyclopédique et critique des publics [Online]. Available at: https://publictionnaire.huma-num.fr/notice/contre-enquetes-osint/.

Deneuville, A., Hernández López, G., Rasmi, J. (2022). Un âge de nouvelles enquêtes ? *Multitudes*, 89, 58–66.

Derrida, J. (1993). *Dissemination*. Athlone Press, London.

Dubberley, S., Koenig, A., Murray, D. (eds) (2020). *Digital Witness: Using Open Source Information for Human Rights Investigation, Documentation, and Accountability*. Oxford University Press, Oxford.

Esquerre, A. (2018). *Le vertige des faits alternatifs*. Textuel, Paris.

Foucault, M. (2007). *Security, Territory, Population: Lectures at the Collège de France, 1977-1978*. Palgrave Macmillan, London.

Foucault, M. (2011). *The Courage of Truth: The Government of Self and Others II – Lectures at the Collège de France, 1983-1984*. Palgrave Macmillan, London.

Foucault, M. (2016). *Discourse and Truth and Parrēsia*. Chicago University Press, Chicago.

Fuller, M. and Weizman, E. (2021). *Investigative Aesthetics: Conflicts and Commons in the Politics of Truth*. Verso Books, London.

Harley, J.B. (1989). Deconstructing the map, cartographica. *The International Journal for Geographic Information and Geovisualization*, 26(2), 1–20.

Heller, C. and Pezzani, L. (2014). Traces liquides : enquête sur la mort de migrants dans la zone-frontière maritime de l'Union européenne. *Revue européenne des migrations internationales*, 30(3–4), 71–107.

Joliveau, T. (2016). Google Maps ! Plus qu'une carte. *La Géographie*, 1560, 14–19.

Joliveau, T., Noucher, M., Roche, S. (2013). La cartographie 2.0, vers une approche critique d'un nouveau régime cartographique. *L'Information géographique*, 77, 29–46.

Kittler, F.A. (1999). *Gramophone, Film, Typewriter*. Standford University Press, Redwood City.

Kittler, F.A. (2015). *Médias optiques. Cours Berlinois 1999*. L'Harmattan, Paris.

Korzybski, A. (1995 [1933]). *Science and Sanity: An Introduction to Non-Aristotelian Systems and General Semantics*. Institute of General Semantics, Lakeville.

Kurgan, L. (2013). *Close Up at a Distance: Mapping, Technology and Politics*. MIT University Press, Boston.

Lacoste, Y. (1976). *La géographie, ça sert, d'abord, à faire la guerre*. Éditions François Maspero, Paris.

Latour, B. (2007). *Petites leçons de sociologie des sciences*. La Découverte, Paris.

Lavoie, V. (ed.) (2017). *La preuve par l'image*. Presses de l'Université du Québec, Montreal.

Le Deuff, O. and Roumanos, R. (eds) (2021). OSINT – Open source intelligence. *I2D – Information, données & documents*, 2021-1.

Ledoux, A. and Zabunyan, D. (eds) (2023). *La "preuve par l'image" : nouvelles pratiques et enjeux contemporains*. Classiques Garnier, Paris.

Lorde, A. (2018). *The Master's Tools Will Never Dismantle the Master's House*. Penguin Books, New York.

McLuhan, M. (1964). *Understanding Media: The Extensions of Man*. McGraw Hill, New York.

Newman, N., Fletcher, R., Eddy, K., Robertson, C.T., Kleis Nielsen, R. (2023). *Digital News Report 2023*. Reuters Institute, Oxford University, Oxford.

Noucher, M. (2017). *Les petites cartes du web. Analyse critique des nouvelles fabriques cartographiques*. Éditions Rue d'Ulm, Paris.

Noucher, M. (2023). *Blancs des cartes et boîtes noires algorithmiques*. CNRS Éditions, Paris.

Parks, L. (2018). *Rethinking Media Coverage: Vertical Mediation and the War on Terror*. Routledge, London.

Pickles, J. (ed.) (1995). *Ground Truth: The Social Implications of Geographic Information Systems*. Guilford Press, New York.

Sarrut, M., Echeverri Zuluaga, J., Valenzuela Amaya, S. (2023). Briser le mythe de la "jungle qui tue" : analyse du rôle des intermédiaires dans la traversée du Darién (frontière Colombie-Panama). *Revue européenne des migrations internationales*, 39, 15–42.

United Nations and University of California, Berkeley (eds) (2022). *Berkeley Protocol on Digital Open Source Investigations: A Practical Guide on the Effective Use of Digital Open Source Information in Investigating Violations of International Criminal, Human Rights and Humanitarian Law*. United Nations Human Rights, Office of the High Commissioner; Human Rights Center, UC Berkeley School of Law, Berkeley.

Weizman, E. (2017). *Forensic Architecture: Violence at the Threshold of Detectability*. Zone Book, London.

Zhao, B. and Sui, D. (2017). True lies in geospatial big data: Detecting location spoofing in social media. *Annals of GIS*, 23(1), 1–14.

Zhao, B., Zhang, S., Xu, C., Sun, Y., Deng, C. (2021). Deep fake geography? When geospatial data encounter artificial intelligence. *Cartography and Geographic Information Science*, 48(4), 338–352.

Zuboff, S. (2018). *The Age of Surveillance Capitalism: The Fight for a Human Future at the New Frontier of Power.* Public Affairs, New York.

Zwer, N. and Rekacewicz, P. (2021). *Cartographie radicale. Explorations.* La Découverte, Paris.

8

Exploring Geospatial Technology Imaginaries

Thierry JOLIVEAU
UMR Environnement-Ville-Société, Université Jean Monnet de Saint-Etienne, France

8.1. Introduction

Digital geospatial technologies are now used on a daily basis and on a large scale across the world through a wide range of technical devices. The most obvious example is the map, whose digital avatars are highly diverse, from countless small maps on the web (Noucher 2017) to the professional platforms of national mapping agencies. These technologies also underpin all geolocation and navigation systems, whether embedded in smartphones, autonomous vehicles or even household robots. As such, geospatial information now permeates nearly all domains of decision-making in contemporary societies. A question that is rarely addressed with these technologies concerns the meaning and representations circulating in society regarding them, beyond their strictly functional dimension.

This chapter seeks to explore a new direction by outlining what could be the imaginary of geospatial techniques in the digital era. To emphasize the exploratory character of this proposal, it takes the form of an inquiry that gradually establishes the concepts to be implemented, the nature of the available source materials and the selection of examples – drawn primarily from the field of fictional imagery related to geolocation – before ultimately engaging with their interpretation.

Geospatial Technologies and Society,
coordinated by Thierry JOLIVEAU and Matthieu NOUCHER.

8.2. From sociotechnical imaginaries to a geo-technical imaginary

Research on the imaginary dimension of technology is longstanding, and approaches vary widely. This is evident in the volume edited by Maestrutti and Kröger (2022), who explicitly use the plural "imaginaries". The theoretical references are strong and widely shared (Simondon 1958; Leroi-Gourhan 1964). The imaginary under consideration is social and collective. Jasonoff and Kim (2015, p. 4) have a global and rather stable definition of sociotechnical imaginaries, which are "collectively shared, institutionally stabilized, and publicly performed visions of desirable futures". For Arruda (2015, p. 129), a sociotechnical imaginary must integrate "the set of images, models and beliefs individuals inherit from participation in society that related to a period of time", and, according to Castoriadis (1975), who does not reduce the imaginary to technology, it must also include the radically new creations constantly produced by society. Musso (2022) argues that "the technical object is formed of an essential duality, functional and fictional: it is function and fiction, instrument and imagination [...]" (p. 79) and:

> The imagination is neither the opposite of the rational nor the real, but rather their necessary complement. The imaginary is not only the product of the imagination, from which it must be distinguished: it is a language made up of narratives, stories, and universes of dynamic forms and images with a certain coherence[1] (p. 80).

Flichy (2001) also considers the imaginary to be an essential element in the study of techniques: "Insofar as technical action, like all human action, cannot exist without taking a symbolic form, we cannot conceive or use a technique without representing it to ourselves" (p. 71). When analyzing the emergence of new information technologies in the 1980s, Scardigli (1992) points out the impossibility of separating the creation and the diffusion of technical inventions. He refers to a "social birth of invention" (Scardigli 1992, p. 38) produced through the joint involvement of inventors, industrial actors, administrative and political institutions, social groups and also users, who may appropriate these technologies in their everyday lives. All of these actors rival each other in their imagination regarding this new technology. For such a technology to exist and deploy, a "network of meanings and actions [...] an intimate association of the symbolic and the operational" (Scardigli 1992, p. 38) must take shape. This network accompanies the entire life cycle of the technical object and constitutes its collective imaginary. For Balandier, modernity has contributed to dissociating the operational and the symbolic. In traditional societies, "rite and tool are intimately linked, the symbolic efficacy of the technical act is associated with its mechanical and physical efficacy" (Chanial and

1 The author translated all original French quotations.

Balandier 1994, p. 124). The anthropologist also stresses the central role taken by information and communication technologies in contemporary societies and confirms the importance of network logic:

> Communication technologies penetrate the most deeply, by their very nature. They are linked by a chain of networks. They occupy private space. Through their omnipresence, they intervene increasingly in the construction of reality and in the relationship to events. They modify both the intellectual and the sensory apprehension of the world; they mediate many social relations, they make possible a remote steering of consciousness, and through images and simulations, they participate in shaping a new imaginary – a technical imaginary. (Balandier 1991)

Having previously proposed, drawing on Balandier's (2001) work, to describe geomatics as a technoscience in the making (Joliveau 2004), it now seems consistent to advance the hypothesis that the geographic dimension of information – and the ways it circulates through digital tools and infrastructures – may give rise to a specific technical imaginary.

8.2.1. *In search of a geospatial technological imaginary*

However, an overview of research on this topic does not appear promising. Studies that address the imaginaries of geospatial technologies in a precise way are rare. A search of various bibliographic databases combining the words imaginary/*imaginaire* with English and French keywords focused on these technologies (geographic data/geodata, GIS, geomatics, etc.) produces few relevant results. As soon as the search is broadened to include data, cartography, space or geography, the number of results increases sharply, but without yielding any references that specifically refer to geospatial techniques. In the scientific literature, the imaginary of geospatial technologies is difficult to isolate from other broader imaginaries: the imaginary of digital technologies in general, the geographical imaginary related to space and places, or the cartographic imaginary linked to orientation, location and representation of space and territories. It is possible that the imaginary dimension attaches more to the concrete technical objects produced with these techniques or that implement them, including immemorial objects such as maps, than to the techniques themselves. As Musso (2009, p. 4) puts it:

> The challenge for individuals and society is to 'socialize' techniques, or even to 'naturalize' them, by constantly manipulating two types of representations: 'social macro-representations', which are grand narratives about society, such as 'information and communication

society', 'knowledge' (or conversely, surveillance), and 'micro-representations', which are fictions and discourses related to each specific technical object.

Three sociotechnical objects linked to geospatial digitalization seem to illustrate this typology: two that are rather "micro" – the smartphone and the autonomous vehicle – and one that is more "macro" – the smart city. Nova (2018) identified six different figures through which users represent and experience their smartphones: the leash, the prosthesis, the mirror, the magic wand, the cocoon and the empty shell. Beyond the device's purely operational functions (communication, information, music, video, etc., and of course geolocation, GPS and mapping), smartphone use carries meaning and imagination. The contemporary autonomous vehicle is another technical object imbued with a strong imaginary dimension. Kröger (2022) retraced its history in the United States, from the first radio-controlled cars of the 1920s and the *Futurama* – a giant diorama built for the 1939 World's Fair with 10,000 remote-controlled cars traveling through a model of a futuristic American landscape – to modern car advertisements and the 2023 experiments with autonomous taxi services in San Francisco. The car that becomes autonomous to the point of escaping its users and turning malevolent and evil is an old theme in science fiction literature and cinema[2], and it continues to shape contemporary imaginaries. However, current autonomous vehicles are fully connected platforms, filled with geospatial technologies enabling them to locate themselves and capture their immediate environment (Wilken and Thomas 2019). This dimension, however, is rarely highlighted in advertising and communications aimed at the general public. The smart city also generates a contrasting imaginary, both as an ideal future and as a dystopian warning, a "liberating or totalitarian city" (Minvielle and Wathelet 2020). It is also a typical example of a global technological narrative, a "self-fulfilling discourse" (Picon 2013) driven by large IT and urban services companies and relayed in planning circles (Söderström et al. 2014). Once again, geospatial technologies are central to this "spatial intelligence" that defines the smart city according to Picon (2015), even if this centrality is seldom emphasized.

These three examples show how geospatial technologies lie at the core of technical innovations that fundamentally reshape how societies relate to space. They contribute to new configurations of presence and absence, new modes of managing distance and new forms of controlling spatial positions through increasingly realistic

2 For example: *Christine* by John Carpenter (1983), adapted from a novel by Stephen King, or *The Car* by Eliott Silverstein (1977).

expressions of co-spatiality. A strong imaginary dimension is therefore attached, not to geospatial technologies themselves, but, on the one hand, to the concrete technical artifacts that individual and collectives use and experience, and, on the other hand, to the more global yet subterranean transformation produced by a generalized geospatial digitization of human action. Then, the question is: Where are these imaginaries of technology forged?

8.2.2. *The different forges of sociotechnical imaginaries*

In his article on the imaginary of Information and Communication Technologies (ICT), Musso (2009) distinguishes – following Flichy (1995) – three different imaginaries:

1) the imaginary of the designers and promoters of a technical innovation, including both local and global companies in the field as well as the individuals who work for them;

2) the imaginary of the major public organizations meant to regulate the domain, which could be called the regulators;

3) the imaginary of what he calls "literary figures", meaning novelists, journalists, filmmakers, comic book authors, etc.

These different imaginaries are, of course, neither sealed nor homogeneous. Representations and discourses around the early Internet varied across project initiators, designers, early adopters and mediators, yet circulated with a certain continuity (Scardigli 1992; Flichy 2001).

Each of these imaginaries is also riddled with contradictions and tensions. This typology, however, requires updating. Journalists and creators of fiction do not operate under the same regime of truthfulness. It is therefore preferable to limit the third category to the press, while adding to it – without conflating them – social networks, which today largely interfere with traditional journalism. A fourth category should gather the imaginary produced by creators of fiction in a broad sense, including literature, cinema, art, video games, etc. Finally, a fifth category needs to be introduced: the imaginary of the users targeted by these technologies, whether they adopt, ignore or reject them, along with their expectations, dreams, hopes, fears and anxieties. This last imaginary feeds on the other four while simultaneously enriching them, as users are the ultimate audience of the discourses produced by all other actors.

In the literature on geomatics, the first imaginary, linked to the designers and promoters of geographic information systems, a new technical object emerging in the late 1980s, was examined early on by Roberts and Schein (1995) in their chapter of *Ground Truth*, a pioneering book on GIS criticism edited by Pickles (1995). Studying the promotional discourse and imagery relating to remote sensing data, GIS software and GPS systems in the United States in the early 1990s, the authors highlighted the imaginary cultivated by companies of the time. According to them, advertising discourse promises through these technologies a complete form of remote appropriation of the Earth, enabling control, management and manipulation not only of data, but, by extension, of the world itself. It conveyed a sense of power of seamlessly flying over the planet's surface, without boundaries or constraints, and appropriating it for ourselves. The authors also note the recurring analogy between operating a computer and driving a car in these advertisements. They interpret this as evidence that, in the imaginary of early promoters of geographic computing, the old Fordist automobile technology – the emblematic shaper of 20th-century space – was being superseded by a new, immaterial, post-Fordist spatial technology[3]. They further read these automotive metaphors as signals of the masculinist orientation of the geomatics industry, geared toward domination and possession, and of the voyeuristic impulse associated with satellite images, which enables us to see without being seen.

Revisiting such analyses 40 years later for companies working on geospatial technologies would be fruitful, revealing both continuities and changes in this imaginary. The media ecosystem, however, would be vastly different: YouTube videos, TED talks, presentations at trade shows, etc., would all require attention. It would also be necessary to investigate the imaginaries of geomatics professionals working in different fields, like urban planning and environmental management, as well as those of geodata scientists, these new players coming from digital geovisualization and coding, and contributors to the OpenStreetMap, some of whom are also geomatics or IT professionals. Another research avenue would be to set up continuous observation of geospatial imaginaries of public institutions, the press and social media, and users themselves. Nova (2014) pointed out that the imaginary of designers, artists and architects often depart from the more conventional technical imaginaries of the future found in science fiction. In the remainder of this chapter, the focus will be placed on the geospatial technical imaginary produced by creators of fiction, since it provides images and narratives that feed all of the other forges of sociotechnical imaginary.

---

3 It was at this point that Al Gore, Vice President of the United States, developed his ambitious Information Superhighway project, modeled on the highway infrastructure of the 1930s.

8.3. Inventorying fictional imagery of geospatial technologies

The aim here is to explore the fictional imagery[4] associated with geospatial techniques, that is, the full range of literary, pictorial, cinematographic and other creative representations, that fiction makers propose to the public, to approach the imaginaries that underlie them. It must be recalled at the outset that there is an extensive and steadily growing body of literature on the presence of geographic techniques in works of fiction. Conley (2007) was the first to publish a monograph dedicated to analyzing the role of maps in films. Castro has also devoted several studies to this topic (Castro 2008, 2011). The relationship between mapping and art has likewise generated numerous publications, among which we may cite Tiberghien (2007) and O'Rourke (2013). There is also a wide range of works on literature (Maleval et al. 2012; Engberg-Pedersen 2017; Collot 2018), as well as research workshops recently organized on this topic (Bischoff et al. 2024; Hopquin and Lemarchand 2024), to mention only France. We should also note the volumes published by the "Collectif Stevenson", which bring together original or unusual maps, sometimes taken from works of fiction and accompanied by scholarly commentary: Mappa-insulae, Mappa-urbis, etc. These works are often fascinating and highly relevant to our inquiry, but they cannot serve as its primary source. First, the technical objects considered are mainly maps, almost always in non-digital form. Second, most studies focus on a single type of media (literature, cinema or art) without attempting to identify possible similarities or commonalities between them. Their principal objective is usually to understand the role that maps play in the composition of a specific work, author or genre, rather than to explore the broader imaginary associated with mapping in fiction. In our investigation, we need a set of diverse examples depicting fictional uses of digital geographic information techniques. This requires identifying corpora that inventory such uses in fiction. Before reviewing several catalogues of fictional uses of geographic tools, we will begin with a brief presentation of a collaborative encyclopedia that compiles spatial tropes.

8.3.1. *A corpus of recurrent spatial narrative patterns in fiction*

The *TV Tropes* website is a US-based wiki-style encyclopedia dedicated to an extremely detailed inventory of tropes, in other words, recurring narrative motifs found in stories and fictional works[5]. Its analysis is of interest here for several reasons. First, the site offers highly precise descriptions of figures appearing in pop culture, covering "classic" fictional works from cinema and literature, as well as

4 The formula is borrowed from Vant (1981), who applied it to a different subject.

5 See: https://tvtropes.org/.

science fiction and fantasy, television shows, video games, anime, manga and so on. Second, the site's classification system, which is based on folksonomy, or spontaneous collaborative categorization, is rich, although complex. Certain trope categories are directly relevant to our concerns: geographic information technologies in the broader sense. Mapping (*Map Tropes*) is of course represented, as are navigation (*Navigation Tropes*), wayfinding (*Pathway Tropes*), hiding (*Stealth Tropes*) and coordinates (*The Longitude Problem*). Other categories relate directly to positioning tools such as *The Magic Compass*, *The Magic Map* and *The Maze*. There are many other categories, some of which are very specific. For each one, the site provides examples drawn from specified works. It is therefore a very useful inventory, impressive for the breadth of its content, which potentially spans every imaginable theme and medium. However, the labyrinthine profusion of its categories makes effective use challenging. Moreover, the site does not provide any categories specifying the digital dimension of these tropes, which limits its usefulness for our purposes. Nonetheless, it offers a twofold contribution to our investigation. First, it highlights the fact that maps are not the only geographical technique mentioned in fiction. It confirms the diversity of spatial relations addressed through fictional imagery: how to visualize space, hide within it, get lost, or find our way. The repetition of these motifs suggests a certain regularity, a kind of loose structure of fictionalities associated with actions of spatial interaction. It then reveals that these "spatial tropes" can often be linked to more abstract tropes. For example, the trope *Map All Along* belongs to the more general trope *Hidden in Plain Sight*, of which Edgar Allan Poe's *The Purloined Letter* is the canonical example. Here, the map is just one of many objects that can express the same symbolic function. What is more, the spatial trope often turns out to be a specification of a more general reference. Hiding in space is a special case of the trope of secrecy or social/informational concealment. Getting lost can mean complete disorientation, loss of memory, loss of reason, or even of life. Works of fiction fully exploit the metaphorical dimension of space. In short, *TV Tropes* is of little help in identifying fictional uses of geospatial digital technologies. Nevertheless, it proves useful for situating the fictional imagery we study in a more global context, one that relates to what Lussault calls the "ordeal of space" (Lussault 2009).

8.3.2. *Inventories of geospatial apparatus in works of fiction*

Two websites have made it their mission to catalog the presence of maps in films[6]. There do not appear to be any such catalogs for other arts and media. On

6 There is also a list of globes in films: https://www.senscritique.com/liste/les_globes_terrestres_dans_les_films_mappemondes/1756789.

The Cine-Tourist[7], Lack (n.d.) identified the presence in films of maps of cities he knew well: Geneva, Paris, London, etc. The site lists 400 films in which maps appear (more than 1,400 in total), each accompanied by a reproduction. Although the site itself offers little commentary, its author has published several books on this subject. The Tumblr-based site *Des plans sur des cartes sur des bandes*[8], created by Klimowski's (n.d.) website, provides more than 1,000 screenshots showing maps in films, each captioned with the film's title. The images are not commented on but are tagged with searchable keywords, for example, #globe #war #united states. These two websites offer a very useful starting point for identifying the existence of a map in a film, but they do not describe how the map functions within the narrative. Furthermore, they focus on maps without distinguishing whether they are digital or not.

A final inventory will serve as our reference corpus. This is the collective website (e)space&fiction[9] hosted on wordpress.com (Joliveau et al. 2009). Its aim is to bring together a curated collection of *geospatial apparatus*, that is, the various tools and instruments used to describe and analyze space. Maps of different kinds are, of course, the most obvious means of capturing the spatial dimension of the world in fiction, but they should not eclipse terrestrial globes, world maps, compasses, theodolites, geographic information systems, postal addresses, street signs, place names and many other geospatial apparatuses. Another distinctive feature of this site, compared to the two previous ones, is its ambition to cover all types of media and genres: cinema, literature, art, comics and so on. Published online as blog posts, the apparatuses are described using a list of keywords that specifies the type of post, the type of work and the technique of the apparatus in question. Among these, some explicitly involve digital technologies.

8.3.3. *The (e)space&fiction website*

In 2024, (e)space&fiction gathered a little over 400 posts of various types (geospatial apparatus, short stories, quotations, etc.), varying in length and analytical depth. Cinema dominates this total (245 posts), far ahead of literature (69), art (40), comics (34) and television (28), with other domains represented only marginally (architecture, design, games, music, theater, etc.). We notice the predominance of moving or still images over text, which can be explained by the tastes of the

---

7 See: https://www.thecinetourist.net/. The site is still online but is no longer updated.

8 See: https://cartographie-et-cinema.tumblr.com/. The site has been discontinued since 2018.

9 See: https://spacefiction.fr/. The site was created in 2009 by Joliveau and Caquard as a collective blog to which several authors contributed, including Mazagol, Tuffery, Noucher and others. It has continued to be updated more sporadically over the past 3–4 years.

contributors, mainly men over 30 with university degrees, and by the disproportionately large number of posts written by only three of them. The sample is therefore not representative but remains of interest due to its size.

Type of geospatial apparatus mentioned	Number of posts	Type of geospatial apparatus mentioned	Number of posts
Maps	274	Sculpture	10
Narrative	72	Geographic Information System	9
Models	42	Coordinate systems	8
Satellite or aerial imagery	28	Sign	6
Painting	27	StreetView	6
GPS	19	Augmented reality	5
3D views	19	Poster	4
Photo	15	Video game	3
Virtual globe	13	World map	3
Navigation system	13	Relief map	2
Spatial analysis	12	Cross-section	2
Terrestrial globe	12	Board game	2
Art Installation	11	Textile	2
Plan	11	Topography	2
Web map service	11	Communication system	1
Video	11	Land art	1

Table 8.1. *The different types of geospatial apparatus mentioned on the (e)space&fiction website (as of May 15, 2024)*

The diversity of post genres means that a strictly quantitative approach to the technical types of geospatial apparatus has limited significance, especially since a single post can describe several apparatus and combine different categories. However, Table 8.1 shows that maps outnumber other apparatus, well ahead of narratives, models and satellite or aerial images. Digital devices such as GPS, virtual globes, GIS, StreetView-type devices, virtual reality or augmented reality are much rarer in fiction. Even if the sample is arbitrary, these figures show that fictional references to geospatial apparatus, maps or others, digital or not, are abundant, which justifies undertaking a closer analysis. It should also be noted that while maps dominate, some of them are produced using digital technologies that give them specific characteristics. They are often dynamic or interactive, for instance. Fictional geospatial imagery would thus extend older cartographic imagery. Furthermore, this cartographic imagery is itself embedded within a broader geographical imagery linked to the places themselves as rendered through the geospatial apparatus. In a specific work, it is not always easy to separate the geographical substance (the connection to the place itself) from its carto-graphic expression (the figures or forms through which that connection is represented).

8.3.4. *From "presage" to usage: geospatial apparatus in cinema*

It must first be noted that, in cinema, fictional geospatial apparatuses have often largely preceded the technical inventions themselves (Caquard 2009). Among these technical foreshadowings or "presages" as we might called them, some belong to the diegesis (the narrative universe) and others do not. In the first case, the devices are operated by the fictional characters themselves, while in the second case, the film uses special effects to simulate operations that would only become possible with the advent of computer technology. The animated maps in the docudramas of the 1910s fall into the second category, where a line is gradually drawn on the image of a map to show an explorer's route. In the movie *M* by Fritz Lang (1931), a famous shot anticipates navigation in Google Earth by combining the superimposition of an aerial photograph and a topographic map, the use of a compass to draw circles of increasing radius around a crime scene and a continuous transition from a zenithal view to an oblique view, then to a sequence filmed by a ground camera. Another example cited by Caquard (2009) is the opening scene of Michael Curtiz's *Casablanca* (1942) with its dizzying plunge through a series of cross-fades from the terrestrial globe to the souk of Casablanca. By contrast, Akira Kurosawa's *High and Low* (1963)[10] incorporates a diegetic technical "presage". The police officers in the film manually assemble a veritable geographic information system to locate a kidnapper, in the very year that the concept and term GIS emerged in Roger Tomlinson's Canadian GIS project. Less than ten years later, Robert Wise's *The Andromeda Strain* (1971), analyzed by Mazagol[11], featured a computer that dominates the struggle of scientists attempting to contain the spread of an extraterrestrial virus. Digital mapping, spatial analysis and computer simulation played a fundamental role in resolving the plot.

---

10 See: https://spacefiction.fr/2010/05/16/gis-and-kurosawa-kurosawa-et-les-sig/.

11 See: https://spacefiction.fr/2015/02/20/le-premier-sig-du-cinema-le-mystere-andromede-first-gis-in-cinema-the-andromeda-strain/.

Everything suggests, then, that some fiction films invent, through moving images, a cinematographic solution to a functional problem that technology would later attempt to solve. Of course, developers of digital geographic information tools did not necessarily draw their inspiration directly from these films, which used special effects or early computer visualizations to offer new filmic experiences[12]. This parallel rather leads us to posit a deeper kinship between cinema images and cartography, which would confirm Castro's (2011) hypothesis about the existence of a cartographic (or spatial) rationale for images: "Cinema and, beyond it, the world of images in general is traversed by a thinking of space that finds its theoretical model in the geographic map". Cinema would invent, with its own means, apparatus that prefigure the tools of modern cartography. However, in doing so, it also captures the technical imagination of its time. The 1930s were the era of mechanization and aeronautics, which made aerial views commonplace in newsreels (Asendorf et al. 2013). Cinema thus created novel techniques of geographical representation in line with the technical imagination of its time. Digital mapping techniques put the user in control, transforming the viewer into the pilot of their own relationship to space, to use the analogy already mentioned by Roberts and Schein (1995).

8.4. Several facets of the fictional imaginary of geospatial technologies

Drawing from the descriptions of geospatial apparatus frequently listed in (e)space&fiction and from the imagery associated with them, let us follow a few leads that may outline an imaginary surrounding the use of geospatial technologies. We have seen that the map, whether digital or not, is the dominant geospatial apparatus in fiction. It would be relevant to study whether their fictional use changes once it becomes digitalized. However, given the number and diversity of maps in works of fiction, such an analysis is beyond the scope of this article. Yet another geospatial apparatus persists after digitalization: geographic coordinates and geolocation more broadly. This narrative motif is somewhat less common than the map, but its forms are rich and varied enough to observe the impact of the emergence of GPS and geospatial techniques on the way fictional characters orient and locate themselves. This is the first theme developed below[13]. We then describe the imaginary that accompanied the emergence of a "geospatial hero" in the

12 The documentary Powers of 10 (Eames and Eames 1977), produced by IBM after the invention of the satellite and the first photographs of the Earth seen from space (Blue Marble 1972), is often cited as a possible direct inspiration for Google Earth.

13 Thank you to all those who took part in the "GPS and geolocation" survey launched on (e)space&fiction in April 2023 (https://spacefiction.fr/2023/04/24/gps-et-geolocalisation-a-travers-la-fiction/).

mid-2000s, before turning our attention to a fictional young man who gets lost in geolocation.

8.4.1. *Geographic coordinates in fiction before and after GPS*

Maritime travelogues are conducive to the mention of geographic positions in the form of coordinates. In *Twenty Thousand Leagues Under the Sea* (1869–1870), Jules Verne cites so many latitude and longitude positions that an interactive map of the journey could be created[14]. This digital reproduction of the imaginary voyage allows modern readers to trace the same route that 19th-century readers could follow with their fingers on a planisphere. However, it is notable that the opening paragraph of the other great maritime adventure novel of the period takes the opposite approach:

> "Squire Trelawney, Dr. Livesey, and the rest of these gentlemen having asked me to write down the whole particulars about Treasure Island, from the beginning to the end, keeping nothing back but the bearings of the island, and that only because there is still treasure not yet lifted," writes the mischievous Robert Louis Stevenson[15].

Between revelation and concealment, coordinates have long excited the imagination of creators. Hergé uses this narrative motif in several albums. In one of Tintin's adventures, *Red Rackham's Treasure* (1945), which makes obvious references to Verne (the submarine) and Stevenson (the treasure island), the coordinates shift from a purely factual mention to the main narrative device. By superimposing the three parchments hidden in the models of the *Unicorn*, Tintin reveals the coordinates of the treasure island. After the confusion between the Paris and Greenwich meridians delays the unsuccessful discovery of the island, the treasure is found by pressing the point corresponding to its location on the terrestrial globe located in the crypt of Marlinspike Hall. The filmmaker Steven Spielberg, a great admirer of Hergé, also uses geographical coordinates as the solution to the mysterious code from space in *Close Encounters of the Third Kind* (1977). Also, in the television series *Breaking Bad* (2008), the narrative again turns on a hidden treasure, the fortune the former chemistry teacher Walter White has amassed by manufacturing and selling methamphetamine. He keeps its location on the lottery

14 See the interactive map created by Viglino (2020) https://macarte.ign.fr/carte/5aea2d2515d03442521ddc91347e8427/Nautilus and the film Globodrome (Wagon 2012), which reproduces *Around the World in 80 Days* on Google Earth.

15 Treasure Island, Robert Louis Stevenson.

ticket where he has written the coordinates as if they were the drawn numbers. The use of geographical coordinates is therefore a trope, a classic narrative motif or a commonplace in books and films. It refers to secrecy and discoveries, to the quest and the find, and evokes children's games (hide-and-seek, treasure hunts, etc.). This essentially playful character of coordinates still must be confirmed by gathering a larger number of examples.

The advent of GPS receivers has the consequence of digitizing coordinates and automating their production. On the one hand, the user loses awareness of coordinates, which are automatically converted into a location on a map to which they will be automatically guided by the navigation system. On the other hand, the spread of GPS comes with a heightened visibility of coordinates, which now appear in plain view on Google Maps or Wikipedia[16]. Fictional works, whether literary or cinematographic, make abundant use of geospatial apparatus based on GPS. One of its earliest fictional appearances is in episode 3 of season 2 of the television crime series *Monk*. It is 2003, and fewer than three years have passed since the degradation of the GPS signal for civilians was stopped, making geolocation usable for guiding cars through city streets. In that episode, the multimillionaire Hammond and his wife are murdered in a dead-end street where their GPS had led them. After investigating, Monk proves that the perpetrator hacked the CD where the navigation software and map data were stored at the time. This possible first misuse of a GPS-based geospatial apparatus serves to disorient a character to kill him. The situation is opposite in *The Ghost Writer*, Roman Polanski's 2010 film adaptation of Robert Harris's 2007 novel *The Ghost*. The main character, known as The Ghost, is the new ghostwriter of a former British prime minister. His predecessor died under suspicious circumstances. The Ghost finds himself in his footsteps, investigating troubling facts related to his employer's past. Thanks to the route stored in the car's GPS on the day his predecessor vanished, he is automatically led to a mysterious place. Both the film and the book are based on the idea that the ghost of a dead man guides the hero along his path. The mundane, everyday experience of any driver negotiating their route with a GPS is transformed into a dark and unsettling scene. Thanks to the machine's memory, The Ghost discovers the place that explains his predecessor's death. It is a beautiful thriller scene in which animated GPS maps and a voice (from beyond the grave?) are skillfully used in the montage to trace this itinerary through the land of the dead. This is not exactly a misuse. The geospatial apparatus performs its usual role: automatically guiding its user to his destination. But in this case, the user is the double of the first, whose fate he will share. Another example of a film in which a GPS device is misused is *Black Box* by Yann Gozlan

16 In the 2010s, geographic coordinates became fashionable. Actress Angelina Jolie had the coordinates of her children's birthplaces or adoption locations tattooed on her shoulder.

(2020). Here again, an important message containing geographical coordinates linked to a suspicious plane crash is sent to Mathieu Vasseur, an investigator at the French Bureau of Investigation and Analysis for Civil Aviation Safety. The message is hidden in a scrambled audio file, which he manages to decipher. Entering the coordinates into a GPS leads him to the place where he meets his death, along with the key to the mystery. All of these diverted uses of GPS end badly. The associated imagery is tragic. The childlike games with coordinates have lost their innocence. The machinic dimension, the digital automatism of GPS and its memory capabilities cast a macabre veil over the experience.

8.4.2. *The errancies of digital navigation in fiction*

What, then, can be said about the fictional imagery that accompanies ordinary, everyday uses of GPS to orient ourselves and to move either materially or virtually? Five works, films or novels, build their plots around characters confronted with navigation interfaces and virtual globes in the midst of familiar daily life, whether the movements involved are real or imagined. In these works, digital geolocation forms a crucial narrative mechanism. Each work would require an in-depth analysis that cannot be developed here due to a lack of space.

The first novel based on the use of GPS and digital geolocation is likely *Spook Country* by William Gibson, published in 2007. Its dense plot is difficult to summarize. The story takes place in February 2006, a time when in-car GPS was becoming common but before smartphones made mobile Internet access easy. Hollis Henry, formerly a well-known rock musician, has become a cultural journalist for a magazine recently created by Hubertus Bigend, a wealthy advertising mogul. Bigend assigns her to investigate the "locative art" movement in Los Angeles, which consists of creating digital artworks anchored to specific locations in augmented reality. During her research, she meets Bobby Chombo, a coding virtuoso who programs the servers needed for such works. Through Bigend, Hollis learns that Chombo is remotely tracking a mysterious shipping container wandering unpredictably across the oceans, attracting the attention of various shadowy groups of spies and traffickers who are themselves under surveillance. Hollis is gradually pulled into this chase for the container, whose ultimate purpose remains unclear but seems connected to the war in Iraq. The novel intertwines the journeys of numerous colorful characters between New York and Los Angeles, all converging on the port of Vancouver. Geospatial technologies permeate the novel: spatial tagging of global shipping movements (p. 34), creation of geolocated artworks (p. 66) viewed with augmented-reality or laptops headsets (p. 122), connected screens in cars equipped with GPS (pp. 91–92), whose signals can be tracked remotely (p. 286), phones equipped with location tracking devices, etc. This is not science fiction. All of the technologies mentioned existed at the time, though some were still experimental or prohibitively expensive. Most characters seem unaware of these innovations or simply accept them. Only two notice and fear them. Bobby Chombo, the IT expert in networks, is obsessed with what he calls the "GPS grid" that covers the world. He materializes it by drawing white-powder lines on the floor wherever he goes, taking care never to sleep twice in the same square (pp. 42, 56). Sick and terrified, Bobby bears badly the central role he plays in the hunt for the container. Hollis, for her part, discovers through Bigend the spread of these technologies with growing unease:

> The war on terror. Were they still calling it that? She'd caught some, she decided: terror. Right here in her hand, in Starbucks, afraid to trust her own phone and the net stretching out from it, strung through those creepy fake trees you saw from highways here, the cellular towers disguised with grotesque faux foliage, Cubist fronds, Art Deco conifers, a thin forest supporting an invisible grid, not unlike the one spread on Bobby's factory floor in flour, chalk, anthrax, baby laxative, whatever it was. The trees triangulated on. The net of telephony, all digitized, and all, she had to suppose, listened to. [...] Maybe now, they already were. Listening to her. She looked up and saw the other customers. [...] None of them, in that moment, looking particularly happy. But none of them, probably, touched in quite the same way by this new bad thing, this shadow, fallen across her (p. 160).

Filled with electronic and informational devices based on computer code, the material world is encompassed by a digital world with which the boundaries have become porous, even imperceptible. Hollis and Bobby's concern about GPS tracking touches on this invisible technical mesh now covering and controlling the physical world. However, this anxiety and mistrust seem to fade in the final resolution.

Jonathan Coe published the novel *The Terrible Privacy of Maxwell Sim* in 2010. He helped write the screenplay for the 2015 film adaptation by Michel Leclerc, whose plot stays very close to the novel. Maxwell Sim, left by his wife and in deep depression, accepts a somewhat ridiculous assignment: driving across the country to promote a revolutionary toothbrush. His car is equipped with a GPS receiver featuring a female voice, which he names Emma. He begins talking to it, then confiding in it, and gradually it becomes a kind of friend. Little by little, he forgets his original mission and veers off course to revisit childhood places. This journey forces painful memories to resurface and worsens his depression. He increasingly identifies with the red arrow on the GPS screen. The dissociation between driver and car is still evident on page 143: "When I pressed the button, anyway, a map appeared. It showed the highway I was driving on, and it showed me – or at least my car – as a little red arrow heading determinedly forward in an eastbound direction". On page 184, the driver identifies himself with the GPS sign. "I glanced down at the map screen. There I still was, a little red arrow luckily making its way along the A461". On page 219, the *I* becomes *we*: "We had left Kendal about ten minutes ago, and were now driving along the A6 in the direction of Penrith". This *we* represents Max and Emma (the GPS) together. As Emma becomes more personalized, Max becomes less so, as if they were exchanging identities. The novel is complex, weaving together multiple plotlines around the existential crisis of a solitary man who is absent to himself and adrift in every sense. Sim identifies with a real navigator, Donald Crowhurst, famous for secretly abandoning a solo race in 1969, while continuing to transmit false positions to make it appear as though he were successfully completing the planned circumnavigation. Crowhurst disappeared at sea, apparently descended into madness and committing suicide rather than confess his deception. Sim finds himself in a similar situation. Having believed he had found in the voice of the GPS a remedy for his loneliness, he is pushed to the brink of suicide when Emma's energy fades and she ultimately abandons him. However, the ending of the novel remains optimistic.

In a similar, but more extreme and darker, vein, Lucie Rico also explores in her novel *GPS* (published in French), a hallucinatory story of geolocation. Ariane, the narrator, is an unemployed local news reporter who is also suffering from depression. Her childhood friend Sandrine asks her to serve as a witness at her engagement ceremony. When Ariane refuses to attend, Sandrine promises that she will come to pick her up. On the appointed day, as Ariane waits for her, she receives a Google Maps link on her smartphone ("which you always call GPS by misuse of language", p. 15), indicating that Sandrine has chosen to share her location with her. To find her, Ariane must use the navigation system and head toward the red circle indicating Sandrine's position, a symbol Ariane quickly comes to identify with her friend. When Sandrine disappears without a trace, Ariane tries to understand what has happened by following the movements of the red dot. The entire story of the novel is devoted to the investigation into Sandrine's disappearance, alongside a renewed reading of their shared past. The plot shares similarities with Coe's novel in its emphasis on the central role of geolocation and its use of geographical disorientation as a metaphor for a more fundamental, psychological or existential disorientation. However, Rico's literary treatment is very different. As in Coe's novel, the narrator recounts her own story; but in *GPS*, the narrative is written in the second person singular. Ariane addresses herself directly, and the reader, constantly interpellated, is invited to take on Ariane's actions, thoughts and phobias. Geospatial tools and their uses are at the core of the story, in which the author subtly disrupts space and time through small but cumulative distortions, of which only a few examples can be noted here. This new way of exploring the physical world transforms perceptual logics. Ariane comes to prefer the digital world to material reality. The fixed, stable and permanent world of Google Earth offers reassurance. "You tell yourself that, as seen on the GPS, the castle is much better than in real life. Why is that? The color spectrum? The fixity of the map?" (p. 33).

> In the GPS, the sun never sets, it never rains, and the wind does not blow. Light spreads evenly over the lawns and streams, the lily ponds are charming, and all spaces are clearly named. The clouds print fascinating animal shapes on the ground. The landscape seems to have bought itself, just like Sandrine, a cosmetic surgery. You appreciate this world, without weather, without constraints, without choice (p. 48).

We learn why this world feels better:

> It is augmented. You grasp places that you would never have known how to connect, and the GPS offers you links to go even deeper into the world. You can buy things from the landscape, have them delivered from the landscape, meet people from the landscape. You have all the keys (p. 48).

But this idyllic moment does not last; the space in which Ariane follows the red dot/Sandrine begins to warp, with coordinates becoming increasingly implausible, such as "19°51'85.6″S 2°17.91″N" (p. 141) as wandering and hallucination intensify. The ending of the novel is open but rather unsettling.

The third example is *The Sweet Escape (Comme un avion)*, a film by Bruno Podalydès released in 2015. Once again, the story centers on a character ill at ease in his own life. Michel's marriage has slipped into apathy; he is bored at work – he creates 3D models for a tech company – and he daydreams about the exploits of the aviator Mermoz and the Aéropostale. When he discovers a self-assembly kayak, he decides to travel alone, paddling all the way to the sea. In reality, he ends up in a cheerful and welcoming inn, a veritable Arcadia, to which he returns again and again. Like Crowhurst in Coe's novel, he sends fake photos to his wife to make her believe that he is progressing on his journey, unaware that the images are geotagged. It is the innkeeper with whom he has an affair, who shows him through a photograph sent by his wife that she too is not where she claims to be. The film is a tender, burlesque and poetic comedy that offers a dreamy and offbeat interpretation of digital geolocation. Although Michel uses paper maps to prepare his trip, once out in the field, he relies on a curious object: a smartphone on which he scrolls with his finger through a hand-drawn watercolor map painted on a paper roll. The kayak and this "paper smartphone" undoubtedly express Michel's quiet longing to escape technology and computing. Unlike Sim and Ariane, whose movements are constrained, Michel yearns to travel and fantasizes about adventure. The tiny, peaceful river he paddles down seems to him evocative of the Amazon. The film, which is sensitive and humorous, thus performs a playful "detournement" of geospatial tools. It also plays on the vaudeville mode to highlight the risk of surveillance, often associated with digital geolocation. The staging of the couple's reunion at the end of the film, with Michel paddling in his kayak while his wife walks along on the riverbank beside him, exchanging glances, remains open-ended. Will their stalled relationship begin to move again? Will they go their separate ways? The doubt remains.

Let us conclude with what may be called a motionless travel experience, which consists of visiting places without physically moving to them by using a virtual globe application. A first literary attempt to capture this experience appears in *Netherland*, a novel by Joseph O'Neill published in 2008, when Google Maps had been in existence for less than three years[17]. The novel offers a disenchanted and melancholy reflection on places and movement, space and separation, and the dissatisfaction inherent to situations of cospatiality where we are physically somewhere yet mentally elsewhere. The novel drifts across multiple locations, without ever settling in any of them. The two main characters are exiles in New York. Hans, the narrator, finds himself isolated in Manhattan after September 11, when his wife has left him to return to London with their son. Lonely and lost in what he calls "his New York confusion", he finds refuge on the cricket field run by Chuck, himself a native from Trinidad, and whose regulars are other exiles. Hans undertakes two virtual journeys, the first from Manhattan to England to the house where his wife and son live (p. 119) and the second from London, where he has returned to live with his family, back to Manhattan and the cricket field (pp. 243–244). He moves to both ends of the zoom in Google Maps but finds no human presence either in the very near or in the very far. Through a simple, precise and poetic style, navigating Google Maps resonates in O'Neill's words exactly like the story of a real journey. He accurately captures the minuscule scale of the virtual travel experience: the coordination of the gaze on the screen and the finger on the touchpad, the motion of the cursor and the succession of colored planes fading into the zoom. With gentle irony, he plays on the parallels between virtual and real travel. Despite the static and artificial quality of Google Maps' world, already noted by Ariane in *GPS*, where the days are always clear and objects frozen as if for

17 See: https://spacefiction.fr/2009/11/01/google-maps-enters-litteraturegoogle-maps-entre-dans-la-litterature/.

eternity, these virtual journeys are, for Hans, a different but nonetheless genuine experience of place – or at least as unreal as the one afforded by physical presence in specific situations. The narrator is constantly eager to apprehend the world through eyesight, yet always uncertain of the reality of what he perceives. In the novel, Google Earth is just one of the tools the humans use in their never-successful attempt to appropriate places by looking at it. Separation makes places and people almost imaginary, as the author writes about his mother who remained in London. The two experiences with Google Earth are rather disappointing for the character. They occur at difficult moments, when Hans suffered abandonment or bereavement. The use of Google Earth is therefore tinged with melancholy and disappointment. Above all, these two scenes stand in symmetrical opposition to another scene of takeoff to observe the Earth, at the very end of the novel (pp. 245–247). The final episode is not virtual; it takes place on the London Eye, aptly named "The Eye". From the wheel, London proves as hard to decipher as the view from Google Earth. But this experience differs also because it is not solitary. Hans shares it with his wife, his son and the other passengers in the cabin, whom he helps somewhat clumsily to identify London's major landmarks. The moment also recalls another elevated viewpoint, this one in New York, and contemplated years earlier with his mother. It closes the novel and completes a broad and subtle palette of relations to places, spanning vision and feelings, presence and absence.

8.4.3. *Jack Bauer, the tragic dimension of the geospatial hero*

While the characters previously discussed were ordinary people who use digital positioning devices in their everyday lives, other fictional figures rely on them to carry out extraordinary activities. The fictional character who most closely embodies the ideal type of the geospatial hero is unquestionably Jack Bauer, the American counter-espionage agent in the series *24*, which was broadcast between 2001 and 2014, precisely during the years when geospatial technologies were expanding. This series is both the seed and a fully developed model for the fictional integration of these techniques in TV series storytelling[18]. *24* is a post-2001 show whose release coincides with Al-Qaeda's attack on the World Trade Center. The hero of the series, Jack Bauer, works for the CTU (Counter Terrorist Unit), a fictional American counter-espionage agency. He strives to thwart the dark conspiracies devised by Machiavellian foreigners with the help of American traitors, each more dangerous and high-ranking than the last. To accomplish this, Jack Bauer systematically uses the GPS signal coupled not to his mobile phone, which did not yet support such

18 See: https://spacefiction.fr/2009/12/02/jack-bauer-geospatial-hero/ and https://mondegeonumerique.wordpress.com/2007/02/10/jack-bauer-heros-geonumerique/.

functionality, but to his PDA (personal digital assistant). At the same time, he remains in touch with CTU headquarters by phone. Jack Bauer and the CTU deploy an entire array of tools for capturing and processing geographic information: maps, aerial imagery, infrared digital sensors, repositionable satellites capable of reading license plates, surveillance cameras and drones. The series anticipates techniques that would only become available later within the field of intelligence and information, such as programmable satellites, spy drones and sensors capable of scanning through walls. The series is also original in its theoretical principle of strict unity. Each season takes place in Los Angeles over the course of a single day; each of the 24 episodes corresponds to one hour in the lives of the protagonists, without ellipses. Jack Bauer's time is the viewer's time. A clock displaying the hour, minute and second appears regularly on screen. Spatial realism arises from this temporal realism. Travel times from one location to another aims to be plausible. If a character needs 30 min to get somewhere, the journey last 30 min for the viewer. Multiple plotlines unfold simultaneously, leading to the frequent use of split screens to display parallel actions. With time running out, characters race around the city in a perpetual fight against the clock, glued to their phones. Even though the series takes some liberties with this spatiotemporal constraint, its near-real-time narrative gives space a palpable thickness. Characters cannot escape the distance. They must cover it continuously. Space is a hindrance, and movement is anything but smooth. *24* does not adopt a model of total remote control and absolute technical mastery. If anything, it would be closer to "*Bug Brother*"! Even the most advanced techniques are fallible. It may take many long minutes to reposition a satellite to track a terrorist. These technologies can also turn out to be useless. After a sophisticated system has been devised to geolocate Jack, the plot may finally be resolved through direct action on the ground. To be effective, the Agency's specialized technologies (satellites, wiretaps and so on) must be connected to the technical infrastructures of society: surveillance camera networks, telephone companies' monitoring systems, etc. They only operate when integrated into regular protocols and devices that can fail, just as the specialized tools themselves. Jack Bauer is a professional operator, who depends on a diversified and fragile sociotechnical network. He is not a technician but a man of action, attentive to concrete situations and always ready to engage directly. He constantly relies on CTU's computer experts (especially Chloe O'Brian) to provide him with data, perform analyses, or put him in touch with other informants. *24* depicts a technical process that is complex, multiple, fragile, fragmented and heterogeneous. Increasingly isolated, eventually cast out from the CTU, which is itself dissolved, Jack Bauer disappears by dissolving into the image of the drone that monitors him. His face breaks apart into a mosaic of pixels before the screen fades to black and the countdown ends. The geospatial hero vanishes from the radar. His electronic echo ceases, his image disintegrates. In reality,

24 anticipates a world that now closely resembles our own. Geospatial tools function as instruments of control and surveillance, both personal and institutional. Pressed by time and mobile in space, modern individuals managed this forced ubiquity just as Jack Bauer did, through technology. With the phone, they can act in several places at once; with GPS, they always know where they are; via their smartphone, they remain permanently reachable. Like Jack Bauer, enslaved by time and imprisoned by space, they attempt to liberate themselves through the snares of technical networks, without this making them any happier or any freer.

8.4.4. *Trapped by geolocation*

This overview of fictional geospatial uses ends with a brief presentation of *Trap Street*, a 2013 film by Chinese director Vivian Qu. The main character, Li Qiuming, is a young surveying apprentice working for a private mapping company. We follow him and his mentor as they survey the rapidly changing streets of Nanjing (Nanking). He shares an apartment with friends and earns a bit of extra money through off-the-books surveillance work for local shops (selling and installing cameras, searching for hidden recording devices). He enjoys playing video games and chatting on his mobile phone. Compared to previous navigation system users, Li is an expert in geospatial tools. During one of his surveying rounds, he meets Guan Lifen, a young woman from a visibly more affluent social background who seems to have her life firmly in hand. She has a job, a car, and wears elegant clothes. Smitten, Li engineers a series of "chance" encounters, and a romance begins to take shape. One night, as heavy rain falls, the surveyors drive Guan back to the subway station, where she forgets a box labeled with her phone number. Li finds the box the next morning and calls the young woman to return it. This fairly standard romantic-comedy premise actually conceals a very original film, both in its portrayal of China in the early 2010s and in the way its narrative structure relies extensively on geospatial techniques. When Li first sees Guan, it is through the viewfinder of his tachymeter, which he rotates to follow her until her silhouette disappears into a side street. The first time he holds her hand, he records the moment with his GPS watch in his map of "favorite places". When the surveyors attempt to survey the Forest Alley where Guan works, the street is rejected by the geographic information system. On the taxi's GPS and the one Guan buys on Li's advice, the entry "Forest Alley" returns "address unknown". Li returns at night to the location, where a plaque indicates the presence of a mysterious Laboratory 203. He records its geographical coordinates with his watch. At work, we overhear a senior cartographer explaining to a junior colleague how to properly create a "trap street", a fictious cartographic element that cartographers add to their maps to detect illegal copies or counterfeits. In this case, Forest Alley is the exact opposite of a "trap street".

It is not a false feature added, but a true feature missing. Laboratory 203 turns out to be a confidential government facility. The box Guan forgot contained USB keys. On his second date with the young woman, Li is kidnapped by intelligence agents who suspect him of having copied and sold the data stored on those keys. The romantic comedy then shifts into a political thriller in which the young man is detained and beaten. He must confess without knowing what he is accused of: "It is up to you to prove your innocence, not to us to prove your guilt", his interrogators say. In any case, proof is irrelevant since "the Services only accuse the guilty". Li is eventually released. The company he worked for loses its mapping license, and he himself is fired.

Shot without authorization and never distributed in China, the movie offers a sharp perspective on the country's transformations during the 2010s: the rapid spread of digital technologies, the emergence of private life, the real-estate speculation causing the destruction of traditional neighborhoods, and the shifting values of younger generations. It also reveals the policing and repressive nature of the Chinese regime, where state power is total and arbitrary. But what interests us here is the role of topography in the film. Guan nicknames Li "the surveyor", the topographer, the one who knows locations and draws the boundaries of physical entities. Li's life, in contrast, unfolds in a fluctuating, disorderly, marginal world, borderline we might say. His whimsy and cheerfulness both charm and unsettle Guan, who lives in a more structured environment and seems to possess the traditional codes of power, whereas Li appears unaware of the invisible, secret, implicit and often arbitrary boundaries imposed by the state. Is this a generational issue? Li's young colleague at the mapping firm claims that she knew the Laboratory 203 was "confidential". Is that certain? Is there in China, on one side, a world governed by cartographic precision and rigor, and, on the other side, a more blurred and indeterminate social and political space where we never quite know if, or when, we have crossed a forbidden line? One new element nevertheless becomes apparent: technical competence. In this respect, Li surpasses the Chinese agents. He is surprised that they did not encrypt their USB keys if they contained confidential data. This is unthinkable to him as an "expert" in surveillance. Could this argument of incompetence and negligence be what leads them to ultimately release him?

Trap Street is fascinating because it narrates, through fiction, the moment when Chinese society confronts the arrival of digital technologies and their use for population surveillance. It interlinks the state, the private sector and ordinary citizens, along with the murky fringes of trafficking and corruption. But technical competence appears, at that time, to lie more with the private sector than with the state and the Communist Party. We know how extraordinarily successful China's private digital giants would soon become, and how President Xi Jinping eventually curtailed their autonomy by repressing their leaders and restoring the Party's control over their companies.

8.5. Findings, limitations, outlook and conclusion

The aim now is to synthesize the main results of the inquiry on what might constitute a geospatial imaginary, briefly outline some of the study's limitations and sketch out some prospects for future research before concluding.

8.5.1. *Key findings*

The largely invisible nature of geospatial technologies does not make it easy to analyze the specific imagery that might be associated with them. Research devoted to the imaginaries of designers, regulators or commentators of these technologies remains rare. Yet, as this chapter has shown, enough material exists to analyze imaginary revealed through the imagery associated with their use in works of fiction.

8.5.1.1. *The imaginary of fictional characters*

As we have shown with respect to cinema (Joliveau and Mazagol 2016), an analysis of fictional imagery is useful in highlighting how geospatial apparatus of all kinds are placed in context and put into action. This agency, this role of operator, is an important function of maps (Jacob 1992; Besse and Tiberghien 2017). Observing these instruments in use, even in a fictional context, is enlightening because the imaginary of technologies is shaped through their practical employment. Scardigli (1992) even emphasizes the importance of the small actors of everyday life, who, through their representations and uses of innovation, manage to "produce meaning". The collected material mainly describes the use of geospatial technologies by characters from films, series and novels. These fictional users, endowed with varying degrees of competence, used these tools in their daily life and express through their acts and words their surprises, questions, enthusiasms and anxieties. These invented stories thus forge a fictional imaginary of geospatial technologies that can touch or unsettle the audience, who itself uses these techniques. It is therefore justified to use the analytical grid proposed by Scardigli to bring out the major lines of this technical imaginary related to usage. The grid is organized around a fundamental relationship to technical progress that endows every innovation with a positive or a negative polarity. Innovation is always perceived, sometimes contradictorily by the same actor, as both miracle and curse, desire and fear, hope and threat, etc. This binary opposition is of course too simple. It merely reproduces the famous ambivalence of technology as *pharmakon* – both remedy and poison – which Plato first developed in the *Phaedrus* about writing. This concept was criticized and attempted to be overcome by Stiegler (Mallet 2019). However, this bivalence remains acceptable as a first approximation. Scardigli proposes distributing this duality across seven major themes or stakes: (1) power; (2) knowledge; (3) memory; (4) social justice; (5) social bond; (6) economic prosperity; and (7) space-time. We will discuss four of these, followed by a fifth more general one.

8.5.1.2. *Space-time*

We might assume that the space-time issue is the one most directly concerning our fictional users. But questions of mobility and speed seem purely functional and of little symbolic significance. The expectations of our characters toward geospatial tools are moreover diverse. Some have not chosen to use them. Monk and The Ghost learn about GPS technology when their stories begin, Michel (*The Sweet Escape*) discovers the possibilities of geolocation of photographs only after the fact. Holly (*Spook Country*) and Sim (*The Very Private Life...*) are required to use GPS for professional purposes; the former with reluctance, the latter asking himself when thinking of the satellites pointed at him: "Was the idea reassuring or worrying? As usual, with the new realities of life, I hesitated" (p. 144). He is quickly seduced, less by geolocation efficiency than by the tone of the guiding voice. Ariane (*GPS*) and Matthieu Vasseur (*Black Box*) seem to use GPS in a mundane way before the event that upends their lives in the novel. Chloe (*24*), Bobby Chombo (*Spook Country*) and Li (*Trap Street*) are professional users of geospatial technologies. Only Li displays a genuine passion for them. In any case, the transformation of social relations to space-time affects the other stakes identified by Scardigli.

8.5.1.3. *Social bond*

What effect does the use of geospatial technologies have on the frequency and intensity of interpersonal interactions? The case of Hans (*Netherland*) is typical of the expected functions of these technologies, which are supposed to eliminate distance. Hans travels through Google Maps to compensate for the frustration of being separated from his wife and son. This experience leaves him with a feeling of unfulfillment and melancholy, even though physical travel also struggles to heal absence. Google Maps therefore does not compensate for Hans' loneliness. Maxwell Sim and Ariane soothe their isolation through GPS, which replaces missing or unsatisfactory social relationships. This exclusive new relationship with the GPS leads them to sever ties with friends or relatives. In this regard, recall that GPS may be considered the first robotic assistant to reach a mass audience. An analysis of discussion forums by Dervin (2012) showed that in the early 2010s, it was common for car GPS users to give their device a name, attribute to it a gender and nationality (by choosing the voice's gender and accent), and even ascribe emotions to it, such as irritation when recalculating after a wrong turn by the driver. For Dervin, GPS must be understood as a "digital individual" in its own right. Jonathan Coe pushes this behavior to the extreme. Sim treats his GPS as a real, albeit imaginary, person and falls in love with it. Ariane's case is different because she does not really use the navigation function to move around and does not activate the guiding voice. She associates the map cursor, which is supposed to indicate her own position, with the position of her friend Sandrine. Through what resembles a hallucinatory process, the GPS marker becomes the avatar of a real person, and its movements on the map are Sandrine's. The GPS allows Ariane to believe that she is still in contact with her friend. This is an allusion, of course, to the myth, but also perhaps to the "breadcrumb trail" ("fil d'Ariane" in French), a feature of computer interfaces that allows a user to retrace navigation on a website. Sandrine has disappeared, and the GPS is Ariane's only means of communicating with her. The symbolism might appear positive if it were not a sign of mental disturbance. Finally, as seen above, the geospatial hero Jack Bauer ends in solitude and exile and at the end of *Trap Street,* Li, destabilized, searches frantically through the hotel room which he shares with Guen to uncover hidden microphones and cameras.

8.5.1.4. *Memory*

Here again, these technologies have an ambivalent effect. They allow everything to be recorded while staying connected, but human memory capacities risk deteriorating as they are delegated to machines. Should these machines fail, the memory would be entirely lost. *24* is a perfect illustration of how geospatial technologies facilitate *synchorization*, the "synchronization-in-space" of daily activities described by Beaude (2012). They offer a shared and durable memory, updated in real time and accessible remotely to signal someone's location, know the location of others, confirm the possibility of co-presence or possibly avoid it if necessary. Memory is central to the use of The Ghost and Mathieu Vasseur. It allows them to retrieve a location stored by someone else. This implicitly raises the question of access to this shared memory. Earlier examples involved storing digital recordings of routes in the vehicle's onboard system. But the generalization of smartphone navigation shifted the storage of data to large digital platforms. This brings us back to the risks of surveillance and violations of privacy highlighted by Holly's and Bobby Chombo's GPS grid. As Desbois (2010, p. 8) notes, in *Spook Country*, the geolocated virtual artworks in Los Angeles often have a memorial function related to loss, mourning and remembrance (the deaths of Francis Scott Fitzgerald, Helmut Newton and River Phoenix, for instance). For him, nostalgia is a classic motif of the digital world. This is confirmed by Hans, Sim and Ariane, who use geospatial devices to revisit their past.

8.5.1.5. *Power*

Digital technologies are often legitimized through the liberation from domestic, logistical and material constraints they promise, and the increase in individual and collective autonomy supposedly offered by robots or computer assistants. But they may also enslave humans to machines or automatic systems governed by indecipherable logics. Almost all of the digital geospatial apparatus in our fictions fundamentally concern power and control: horizontal control over our own life and that of others, and vertical, collective and overbearing control. Hans and Ariane's concrete experience with Google Earth is worlds apart from the sense of mastery and power promised by early remote-sensing software advertisers cited by Roberts and Schein (1995). Sim and Ariane lose control of their own reason at one point; Michel and his wife realize that they can mutually surveil their movements. Hollis and Bobby Chombo are themselves watched by unknown networks with unknown aims. Jack Bauer continually battles hidden powers whose technological capabilities surpass his own and which he can resist only temporarily through courage and inventiveness in the field. *Trap Street* directly raises the political question of the democratic issues tied to the development of geospatial technologies.

8.5.1.6. *Happiness or misfortune*

This category subsumes all of the headings in Scardigli's (1992) grid. Technological imaginary typically combines promises of happiness with predictions of misfortune. The fictional imagery of geospatial technologies emerging from our inquiry does not present this duality. Its dominant tone is troubling, dark, even ill-fated, as if their use led almost inevitably to troubles and dereliction. The contexts in which these digital geolocation tools are used are often tragic: murder, espionage and surveillance. Geospatial technologies seem to be more associated with drama than comedy. Most characters who use these tools, when we encounter them, are, to varying degrees, disoriented, maladjusted, lost, unhappy, depressed, lonely, etc. Their use of technology mainly generates anxiety, dissociation, paranoia, hallucinations or boredom at best. Only Li from *Trap Street*, the youngest character, escapes this depressive profile. He is cheerful and carefree, and his use of geospatial apparatus is playful and conscious, until things go wrong. What becomes of these characters next? Two achieve a measure of serenity: Hans and Sim. Michel's case is less clear. Li appears deeply shaken by his ordeal. The ending of *Spook Country* says nothing on the future of Hollis' and Bobby's anxiety. Ariane's and Jack Bauer's destinies are clearly tragic.

The imaginary associated with digital geolocation seems relatively homogeneous and significantly darker than that associated with simple geographic coordinates.

Does using an automatic algorithmic tool produce a feeling of *Great Loss of Meaning*, to borrow Lieutaghi's French play on word GPS, *Grande Perte de Sens*? He writes:

> Do we know more about who we are as GPS refines where we are? It says nothing of true space but substitutes itself for the thought of territory, synthesizes it, breaks it into pre-decision elements, manufactures by itself the decision that it claims to facilitate [...]. It is no longer a piece of the world that we hold in our hands, but the scheme of a predefined mode, to which nothing of ourselves has contributed. (Lieutaghi 2020)

When these automatic navigation systems appeared, geographers often speculated on the loss of conscious connection to space and of spatial knowledge. As if, among the three meanings in French of the word *sens* (sense) identified by Berque (2014) to define spatiality: "physical orientation in space-time, capacity for sensation in our flesh, and mental meaning" – delegating the first to a machine would severely disturb the other two. The unease may stem from the user's feeling that by optimizing their decision of location, they are simultaneously abandoning two things. They must delegate their choices and their journey to an automatic device. Also, they must accept that they "inhabit the geocoded space [...] entirely marked and measured, [...] which, to a great extent, guides and controls us" (Desbois 2010, p. 193), symbolized by the grid in *Spook Country*. Several works (*The Terrible Privacy…*, *GPS*, *Netherland*) show that the characters are fully aware of the technical systems (antenna, network, satellite, etc.), enabling constant localization, and that this awareness heightens their anxiety. Perhaps what emerges here is an imaginary larger than that linked to the use of a technical device. This imaginary would unfold at the macro level discussed by Musso (2009), wherein geospatial digitalization appears as a threat to individual freedom and autonomy.

8.5.2. *Limitations and prospects*

Our inquiry obviously presents certain limitations that further research will need to address.

8.5.2.1. *Broadening the angle and scope of the study*

The study focused on the imaginary associated with digital geolocation technologies, while maps, whether digital or not, are the most common fictional geospatial apparatus. For this reason, we set them aside. An analysis of how the digitization of maps affects their fictional uses remains to be undertaken. Furthermore, our sample of examples centered on ordinary and everyday uses of geolocation. To explore its military, security or political uses, the sample must be expanded to include works of war and science fiction. For example, in his 2019 novel *Les Furtifs*, French novelist Alain Damasio imagines a near-future universe tinged with fantasy. He anticipates new applications of geospatial technologies for police surveillance and military repression, and he envisions new forms of social and political resistance. A thorough examination of this novel, among others, would shed new light on our survey.

8.5.2.2. *The rhetorical nature of fictional geospatial apparatus*

Another dimension to be explored is the rhetorical nature of fictional geographic apparatus, whether digital or not. As the TV Tropes website illustrates, fiction is characterized by the repetition of more or less conventional spatial narrative (tropes). This does not call into question the value of close and subtle analyses that aim to understand the specific role of geographic apparatus in a work. It does, however, require situating those analyses against a background of shared patterns that circulate consciously or unconsciously and can be compared or related to one another. This metaphorical dimension of spatial references must also be part of the analysis. French expressions such as *être déboussolé*, *perdre le Nord*, or *être à l'Ouest* or, in English, "to be off the map", "to lose one's bearings", "get lost!" do not refer to geographical (dis)orientation. Purely metaphorical uses of a fictional geospatial apparatus appear to be rare, but this broader, psychological or existential, dimension frequently resonates in the background of its narrative role. Characters may be lost or disoriented in space, just as they are in their own lives. Gibson articulates this perfectly in *Spook County* (p. 150) when he talks about "the map-lost as well as the life-lost". It would be worth examining whether the spread of geospatial digital technologies has been accompanied by an increase in metaphorical spatial expressions, which might be interpreted as another sign of the "Great Loss of Meaning" mentioned above.

8.5.2.3. *From the characters' geospatial imaginary to the authors'*

Some may find an approach that seems to take at face value the "testimonies" of fictional characters regarding their experience as users of geospatial technologies naive. While this perspective resembles the use of literature by "humanistic" geography (Desbois and Gervais-Lambony 2017), itself aligned with postmodern critics such as Jameson (2007), it assumes that the role of geographical techniques in fiction retains a certain specificity. We could only touch the existential or psychological reality of certain characters whose experience of geospatial technologies is more subtle and nuanced than what has been reported here. A closer reading of several of the works mentioned would be necessary. We might also think that these fictional uses of geospatial apparatus stem primarily from the idiosyncratic imagination of individual authors. However, they often echo or amplify technical imaginaries that are already socially embedded. Fiction often twists or pushes common practices to extremes. It can reveal the fears and hopes of actual users or anticipate original practices. Even when excluding works of pure fantasy and science fiction, as we have done here, fictional geospatial apparatus may be used with great freedom and their functioning may diverge sharply from reality. This imaginative dimension is precisely what interests us. Observing these altered or atypical uses can also help distance from common practices and contribute to a deeper critique of geospatial technologies.

It would be illuminating to explore how authors themselves interpret real-world technological uses and to ask them how they have transposed and embodied these uses within their fictional characters. An analysis of this metadiscourse, through interviews or reflective texts, would constitute a second level of interpretation. An author's works often exhibit a thematic coherence that is essential to understand when interpreting any individual piece. For example, Gibson, the author of *Spook Country*, is renowned for having invented, described and named in 1984, in *Neuromancer*, his cyberpunk science-fiction novel, a new space produced by digital simulation, a "consensual hallucination" that the global connection of computers would transform into a permanently accessible virtual world. His "cyberspace" profoundly shaped the digital imaginary, providing a clear and concise representation, embraced even by its developers and promoters, of what the Internet would become ten years later (Neuromancer, p. 64). The nature of this computer-generated artificial space has been the subject of extensive theoretical debate, even though the term itself is now only used in the fields of defense, intelligence and geopolitics (Douzet 2014). From the beginning of his career, Gibson has shown a sustained interest in the geographical dimension of digital technologies. In 2010, three years after the publication of *Spook Country*, Gibson noted in an opinion column of the New York Times[19] that cyberspace had been surpassed and dissolved: "Cyberspace, not so long ago, was a specific elsewhere, one we visited periodically, peering into it from the familiar physical world. Now cyberspace has everted. Turned itself inside out. Colonized the physical". *Spook Country* recounts this hybridization between cyberspace and physical space, brought about by the advent of so-called ambient or ubiquitous computing, in which devices are scattered throughout the environment. Gibson extends this theme in his next novel, *Zero History* (2010). For Desbois, a close reader of Gibson and cyberspace, Gibson's novels contribute to constructing the imaginary of this emerging hybrid space (Desbois 2011). What our inquiry reveals is that what lies at the heart of Gibson's work according to Desbois (2017): "computing as a source of anxiety and

19 See: http://www.nytimes.com/2010/09/01/opinion/01gibson.html.

as a privileged medium for the spread of paranoia", also permeates more recent novels belonging to quite different genres.

8.5.3. *Conclusion*

The geospatial digitalization of the world is a general phenomenon affecting all dimensions and sectors of society, but it remains subterranean, mediated by a multitude of different digital devices. Geospatial technologies are often incorporated into compact, multifunctional objects that are difficult to "open", such as the smart city, the autonomous vehicle, or the smartphone, which themselves carry a rich and powerful imagery. Even when limiting ourselves to fictional imagery, the categories needed to analyze the types and forms of geographic and cartographic imaginaries are numerous. They combine not only tropes, clichés and genre or media-specific conventions, but also the unique and original inventions of authors striving to represent spatial experience in their creations. This imaginary is generally dark and unsettling. It expresses unease, even a broader sense of dismay and dispossession, associated with feelings of loss and disorientation. This initial overview of fictional geospatial imaginary will need to be expanded by gathering additional examples. Works and authors must be examined in greater detail to better characterize this geospatial imaginary, to assess whether its apparent coherence is real and to understand how it evolves over time.

This inquiry invites further exploration of the possibilities that Balandier foresaw:

> On the side of the techno-imaginary, i.e. an imaginary that no longer functions in 'the old-fashioned way', but [...] by making room for techniques and knowing how to control it for a conscious, informed use of what is at our disposal. (Chanial and Balandier 1994, p. 124)

For "the 'alliance' with technology is never self-evident; it must be continuously negotiated as technology evolves. It requires less fascinated, blinded users than enlightened, vigilant, and critical subjects" (Balandier 1991, p. 10). Developing research on geospatial imaginaries would help to combine criticism and vigilance in the analysis of the future of these technologies.

8.6. Acknowledgments

Thanks to Véronique André-Lamat and Henri Desbois for their valuable proofreading.

8.7. References

Arruda, A. (2015). Image, social imaginary and social representations. In *The Cambridge Handbook of Social Representations, Cambridge Handbooks in Psychology*, Andreouli, E., Gaskell, G., Sammut, G., Valsiner, J. (eds). Cambridge University Press, Cambridge.

Asendorf, C., Gervais, T., Jaumouillé, L., Seban, A., Le Bon, L. (2013). *Vues d'en haut.* Centre Pompidou-Metz, Metz.

Balandier, G. (1991). La technique en jeu : technophiles et technophobes. *Revue européenne des sciences sociales* 29, 5–10.

Balandier, G. (2001). *Le Grand Système.* Fayard, Paris.

Beaude, B. (2012). *Internet, changer l'espace, changer la société.* FYP éditions. Limoges.

Berque, A. (2014). Préface à Pour un vocabulaire de la spatialité japonaise. *Mésologiques. Études des milieux* [Online]. Available at: http://ecoumene.blogspot.fr/2014/04/preface-pour-un-vocabulaire-de-la.html.

Besse, J.-M. and Tiberghien, G.A. (2017). *Opérations cartographiques.* Actes Sud/École nationale supérieure de paysage, Arles.

Bischoff, L., Luis, R., Nègre, J. (eds) (2024). *Écrire avec les cartes.* ENS de Lyon, Lyon.

Caquard, S. (2009). Foreshadowing contemporary digital cartography: A historical review of cinematic maps in films. *The Cartographic Journal*, 46, 46–55.

Castoriadis, C. (1975). *L'institution imaginaire de la société.* Le Seuil, Paris.

Castro, T. (2008). Les cartes vues à travers le cinéma. *Textimage,* N°2 Cartes et Plans.

Castro, T. (2011). *La pensée cartographique des images : cinéma et culture visuelle*. Aléas, Lyon.

Chanial, P. and Balandier, G. (1994). Entretien avec Georges Balandier. *Quaderni*, 23, 119–132. doi: 10.3406/quad.1994.1276.

Collot, M. (2018). Usages littéraires de la carte. In *Cartographier : regards croisés sur les pratiques littéraires et philosophiques contemporaines, collection générale*, Ost, I. (ed.). Presses de l'Université Saint-Louis, Brussels. doi: 10.4000/books.pusl.4484.

Conley, T. (2007). *Cartographic Cinema.* University of Minnesota Press, Minneapolis/ London.

Dervin, F. (2012). Le GPS a-t-il une identité ? L'"object turn" dans l'étude des constructions identitaires. Revue du GERFLINT, Synergies Pays Riverains de la Baltique, 9 [Online]. Available at: https://gerflint.fr/Base/Baltique9/baltique9.html.

Desbois, H. (2010). Code source de William Gibson et les imaginations géographiques à l'ère du GPS. *Géographie et cultures*, 75, 189–206.

Desbois, H. (2011). Le Cyberespace. Retour sur un imaginaire géographique. Carnets de géographes, 2 [Online]. Available at: http://www.carnetsdegeographes.org/carnets_recherches/rech_02_02_Desbois.php.

Desbois, H. (2017). Des cités hallucinées à la grille invisible, William Gibson et les métamorphoses du cyberespace. *Imaginaire informatique et science-fiction*, 10. doi: 10.4000/resf.1043.

Desbois, H. and Gervais-Lambony, P. (2017). *"Les lieux que nous avons connus..." : deux essais sur la géographie, l'humain et la littérature*. Presses universitaires de Paris Nanterre, Nanterre.

Douzet, F. (2014). La géopolitique pour comprendre le cyberespace. *Hérodote*, 152–153, 3–21. doi: 10.3917/her.152.0003.

Engberg-Pedersen, A. (ed.) (2017). *Literature and Cartography: Theories, Histories, Genres*. MIT Press, Cambridge, MA.

Flichy, P. (1995). *L'innovation technique : récents développements en sciences sociales, vers une nouvelle théorie de l'innovation*. Éditions la Découverte, Paris.

Flichy, P. (2001). La place de l'imaginaire dans l'action technique. Le cas de l'internet. *Réseaux*, 109, 52–73.

Hopquin, R. and Lemarchand, F. (2024). Quand la carte désoriente : le dévoiement de la carte dans les arts et les lettres. Laboratoire LASLAR, UR 4256, Caen.

Jacob, C. (1992). *L'empire des cartes : approche théorique de la cartographie à travers l'histoire*. Albin Michel, Paris.

Jameson, F. (2007). *La totalité comme complot : conspiration et paranoïa dans l'imaginaire contemporain*. Les Prairies Ordinaires, Paris.

Jasanoff, S. and Kim, S.-H. (2015). *Dreamscapes of Modernity: Sociotechnical Imaginaries and the Fabrication of Power*. University of Chicago Press, Chicago.

Joliveau, T. (2004). Géomatique et gestion environnementale du territoire. Recherches sur un usage géographique des SIG. Thesis, Université de Rouen, Rouen [Online]. Available at: https://hal.science/tel-04559188.

Joliveau, T. and Mazagol, P.-O. (2016). Les géodispositifs cinématographiques : l'espace mis en action. *Mappemonde* [Online]. Available at: http://mappemonde.mgm.fr/118as4/.

Joliveau, T., Caquard, S., Mazagol, P.O., Tuffery, C., Dubreuil, M., Noucher, M., Saglio, B. (2009). (e)space & fiction [Online]. Available at: https://spacefiction.fr/.

Klimowski, T. (n.d.). Des plans sur des cartes sur des bandes. La cartographie au cinéma/maps, cartography in films [Online]. Available at: https://cartographie-et-cinema.tumblr.com/.

Kröger, F. (2022). L'histoire techno-imaginaire du véhicule autonome. In *Les imaginaires et les techniques*, Kröger, F. and Maestrutti, M. (eds). Mines ParisTech, Paris.

Lack, R.-F. (n.d.). The Cine-Tourist [Online]. Available at: https://www.thecinetourist.net/.

Leroi-Gourhan, A. (1964). *Le geste et la parole. Technique et langage.* Albin Michel, Paris.

Lieutaghi, P. (2020). *La surexplication du monde : un aide-mémoire pour les Temps d'Après.* Actes Sud, Paris.

Lussault, M. (2009). *De la lutte des classes à la lutte des places.* B. Grasset, Paris.

Maestrutti, M. and Kröger, F. (2022). *Les imaginaires et les techniques.* Mines ParisTech, Paris.

Maleval, V., Picker, M., Gabaude, F. (2012). *Géographie poétique et cartographie littéraire.* Presses Universitaires de Limoge, Limoges.

Mallet, A. (2019). Platon et l'écriture, Bernard Stiegler et les algorithmes. *Communication aux Rencontres Philosophiques Clermontoises* [Online]. Available at: http://www.philo63.org/pages/sujets-traites/bernard-stiegler.html.

Minvielle, N. and Wathelet, O. (2020). "Smart city" : d'autres imaginaires existent pour une ville intelligente. *The Conversation* [Online]. Available at: http://theconversation.com/smart-city-dautres-imaginaires-existent-pour-une-ville-intelligente-133264.

Musso, P. (2009). Usages et imaginaires des TIC. In *L'évolution des cultures numériques*, Licoppe, C. (ed.). FYP, Paris, 201–210.

Musso, P. (2022). Techno-imaginaire des réseaux. In *Les imaginaires et les techniques*, Maestrutti, M. and Kröger, F. (eds). Mines ParisTech, Paris, 79–91.

Noucher, M. (2017). *Les Petites Cartes du web. Analyse critique des nouvelles fabriques cartographiques*. Éditions Rue d'Ulm, Paris.

Nova, N. (2014). *Futurs ? La panne des imaginaires technologiques.* Les Moutons électriques, Montélimar.

Nova, N. (2018). Figures mobiles : une anthropologie du smartphone. Thesis, Université de Genève [Online]. Available at: https://archive-ouverte.unige.ch/unige:107645.

O'Rourke, K. (2013). *Walking and Mapping: Artists as Cartographers.* MIT Press, Cambridge, MA.

Pickles, J. (1995). *Ground Truth. The Social Implications of Geographic Information Systems.* The Guilford Press, New York.

Picon, A. (2013). *Smart Cities : théorie et critique d'un idéal auto-réalisateur.* Éditions B2, Paris.

Picon, A. (2015). *Smart Cities: A Spatialised Intelligence.* Wiley, Hoboken, NJ.

Roberts, S.M. and Schein, R.H. (1995). Earth shettering. Global imagery in GIS. In *Ground Truth: The Social Implications of Geographic Information Systems*, Pickles, J. (ed.). Guilford Press, New York.

Scardigli, V. (1992). *Les sens de la technique.* Presses Universitaires de France, Paris.

Simondon, G. (1958). *Du mode d'existence des objets techniques.* Aubier, Paris.

Söderström, O., Paasche, T., Klauser, F. (2014). Smart cities as corporate storytelling. *City*, 18, 307–320. doi: 10.1080/13604813.2014.906716.

Tiberghien, G.A. (2007). *Finis terrae : imaginaires et imaginations cartographiques*. Bayard, Paris.

Vant, A. (1981). *Imagerie et urbanisation : recherches sur l'exemple stéphanois.* Centre d'études foréziennes, Saint-Étienne.

Viglino, J.-M. (2020). Macarte Tour du monde en 80 jours. HackMD [Online]. Available at: https://hackmd.io/@viglino/80jours.

Wagon, G. (2012). Globodrome. Film. 24/30 I/S revue 24 images, Montreal, Quebec [Online]. Available at: https://vimeo.com/51278797.

Wilken, R. and Thomas, J. (2019). Cars and contemporary communication. Maps and the autonomous vehicle as a communication platform. *International Journal of Communication* 13, 25.

8.8. Works analyzed or mentioned

24, TV series. Creators: Joel Surnow and Robert Cochran, 2001–2014.

Black Box (Boîte noire), film. Director: Yann Gozlan, 2020.

Breaking Bad, TV series. Creator: Vince Gilligan, 2008.

Casablanca, film. Director: Michael Curtiz, 1942.

Close Encounters of the Third Kind, film. Director: Steven Spielberg, 1977.

GPS, novel. Author: Lucie Rico, POL, 2022.

High and Low, film. Director: Akira Kurosawa, 1963.

M, film. Director: Fritz Lang, 1931.

Monk, TV series. Creator: Andy Breckman, 2002–2009

Netherland, novel. Author: Joseph O'Neill, Harper Perennial, 2008.

Powers of 10, documentary film, Directors: Charles and Ray Eames, 1977.

Red Rackham's Treasure, comic. Author: Hergé. Casterman, 1945.

Spook Country, novel. Author: William Gibson, Penguin, 2007.

The Andromeda Strain, film. Director: Robert Wise, 1971.

The Ghost, novel. Author: Robert Harris, Hutchinson, 2007.

The Ghost Writer, film. Director: Roman Polanski, 2010.

The Sweet Escape (Comme un avion), film. Director: Bruno Podalydès, 2015.

The Terrible Privacy of Maxwell Sim, novel. Author: Jonathan Coe, Penguin, 2010.

The Very Private Life of Mr Sim (La Vie très privée de Monsieur Sim), film. Director: Michel Leclerc, 2015.

Trap Street, film. Director: Vivian Qu, 2013.

Treasure Island, novel. Author: Robert Louis Stevenson, 1881–1882.

Twenty Thousand Leagues Under the Sea, novel. Author: Jules Verne, 1869–1870.

Conclusion

Escaping Geospatial Somnambulism Through Criticism

Thierry JOLIVEAU[1] and Matthieu NOUCHER[2]

[1]UMR Environnement-Ville-Société, Université Jean Monnet de Saint-Etienne, France

[2]CNRS , UMR PASSAGES, Pessac, France

As noted in the Introduction, the technical process of geodigitalization of the world began 60 years ago with the emergence of early geographic information technology. Over the past 20 years, this process has accelerated through the transcription into digital data of nearly all objects: beings, phenomena, devices, activities, images, even fictional works, etc., that can be located on the Earth's surface. Geographic information science has been at the core of this process of putting the world into geospatial data in all dimensions: theoretical and conceptual invention, methodological innovation, software development, sociotechnical realization, creation of training courses and degrees, publication of scientific journals, dedicated media and conferences, dissemination and popularization, etc. This has been happening in industry and services, government and local authorities, universities and schools. However, geographic information technologies now appear to be dissolving into the broader IT revolution, overshadowed by devices that are ubiquitous, opaque and invisible. The diversification of skills and rapid transformation of professional roles, combined with the massification and banalization of geospatial practices have diluted geomatics; it is everywhere, yet invisible.

C.1. Taking the geodigitalization of the world seriously

This invisibility does not only concern geospatial technologies. As early as the 1950s, Simondon argued that the most powerful source of alienation in the contemporary world stemmed from "misunderstanding of the machine, which is not an alienation caused by the machine, but by ignorance of its nature and essence, by its absence from the world of meanings, and by its omission from the table of values and concepts that are part of culture" (Simondon 1958, pp. 10–11). Geospatial artifacts obviously suffer from this lack of recognition. One way to take the ongoing process of geodigitalization seriously is therefore to fill this gap through renewed and updated forms of knowledge.

This volume seeks to contribute to such an update by providing insights on different fields of activity and important questions in geomatics and geographic information sciences. Without claiming to be exhaustive, its eight chapters help to better understand the diffuse and multifaceted process of geodigitalization. They address issues related to the categorization of geographic data (Territutea Quesnot), the evolution of its uses and representations (Boris Mericksay, Juliette Davret, Grégoire Feyt and Thierry Joliveau) and the extension of its collection and analysis practices into new fields (Allan Deneuville, Juliette Morel and Baptiste Hautdidier). These various contributions raise theoretical, methodological and epistemological questions that call for further exploration. Taken together, they outline a dense and diverse research agenda.

Another philosopher, Winner, encourages us to go further with his idea of "technological somnambulism". Writing at the dawn of the personal computer era, he notes that humans willingly walk with their eyes closed, failing to see that "as technologies are being built and put to use, significant alterations in patterns of human activity and human institutions are already taking place. New worlds are being made" (Winner 1983, p. 108). He goes on to say that "as they become woven into the texture of everyday existence, the devices, techniques, and systems we adopt shed their tool-like qualities to become part of our very humanity" (Winner 1983, p. 108). Technologies thus become "forms of life" (Winner 1983). And the main question to ask is: "As we 'make things work', what kind of *world* are we making?" (Winner 1983, p. 112).

Several specific features of geospatial techniques reinforce their entanglement with everyday life. First is their computational nature, which Ellul described a long time ago as a "flexible, informal, purely technical, immediate, and universal link between technical subsystems" (Ellul 1977, p. 112). Second is the fact that they concern space (and time), which are, to put it in simplistic Kantian terms, a priori forms of sensibility and dimensions of life that are simultaneously completely abstract and self-evident. It contributes to their continuous, ubiquitous, and therefore imperceptible presence. Geospatial techniques also inherit the representational functions of maps, which give them obviousness and elicit the instinctive trust of users turned into spectators. Christian Jacob had already noted that this passive attitude toward maps is a form of "civic obedience" (Jacob 2006, p. 354). The algorithmic opacity that characterizes the contemporary digital spatial technologies amplifies this effect, reinforcing their aura of unquestionable authority (Noucher 2023).

C.2. Developing and generalizing critical approaches

Since the emergence of GIS outside research centers in the late 1980s, scholars have interrogated their epistemological and political nature. This questioning has taken the form of what the scientific literature now refers to as *critical approaches to GIS*. In a seminal article, Schuurman (2000) identified three successive waves of criticism of GIS between 1990 and 2000, followed by an abundance of publications too numerous to detail here. These publications opened up several lines of reflection on GIS: the theoretical and ontological foundations of GIS (Schuurman 2006), the critique of their quantitative assumptions (Pavlovskaya 2006), the development of feminist practices (Kwan 2002; Pavlovskaya and Martin 2007) and the importance of taking a critical look at their history and cartography (Crampton 2010). By 2012, a group of authors expressed considerable optimism about the future of critical GIS (Elwood et al. 2011), but soon after, Goodchild (2014) voiced concerns about a decline in critical scholarship within geomatics regarding Big Data, pointing to a resurgence of digital neopositivism. Since then, research in critical GIS has continued, though it still represents only a small fraction of the broader GIScience literature. To cite a few recent examples, Thatcher et al. (2016) draw conclusions from the 2014 Friday Harbor meeting, held twenty years after the foundational symposium of critical GIS. Noucher (2018) proposed a renewal of critique centered on digital traces as indicators of how geospatial information circulates across the Web. Pavlovskaya (2018) defended the idea that GIS can function as tools of social transformation, capable of reshaping the social imagination toward non-hierarchical class, gender and race relations. Mahmoudi (2020) examined how critical GIS must continually adapt to evolving technological, social, political and economic configurations. We must also note Wilson's volume, devoted to tracing "the new lines related to critical GIS and map unease" (Wilson 2017), as well as the more recent special issue of ACME dedicated entirely to critical GIS (Mahmoudi and Shelton 2022). In this issue, Elwood announces the emergence of a fourth, very radical wave of critical GIS that will contribute to "a conception of worlds beyond racial capitalism, settler colonialism, and cis-heteronormativity as interdependent structures and relations that overdetermine white supremacy" (Elwood 2022, p. 442).

It is essential to underscore that studies do not all share the same vision of what critical GIS should be. Some focus on the technical shortcomings inherent to geospatial technologies: semiological problems, insufficient open-access data, poor interoperability, complex interfaces and so on. Others raise more structural concerns about geospatial innovations. They point to the military and corporate control over geospatial technical development; the expanding role of surveillance in spatial applications; the growing environmental footprint of geospatial data collection, storage and processing; the persistent inequalities in access to data and skills; and the gendered and social biases in the geospatial depiction of the world. A further strand of research aims to "do Critical GIS" (Mahmoudi and Shelton 2022), advocating the use of geospatial technologies to support oppressed and discriminated groups in struggles against injustice and inequality.

Research in critical GIS has been conducted primarily by geographers working in North American universities. The long history and substantial volume of GIS research in the United States, together with the field's early and enduring ties to American industry, which has historically dominated geomatics, help explain the persistence of this minority yet influential research tradition. However, in French geomatics, fundamental and epistemological criticism of GIS remained rare once the initial resistance among some quantitative geographers had faded. Staszak (2001) warned of the risk that geography might become merely a "technique for manipulating data". Lussault offered a more far-reaching critique, arguing that mapping tools in general, whether digital or not, have become ineffective in accounting for the spatial aspects of human action. According to him, there is an unavoidable gap "between the positional space of engineering and the relational space of experience" (Lussault 2009, p. 72). Although this fundamental suspicion was never expressed with the same intensity as in Anglophone literature, most French geographers continued to view geospatial technologies as a strictly technical, ancillary domain, undeniably useful for students' career prospects, but not a fully legitimate object of geographical research (Joliveau 2004). Sociocultural differences also play a role. For instance, political cultures of local governance and planning differ strongly between France and the United States. Between 1990 and 2010 in North America, local public participation was a central arena for critical GIS. In contrast, during the same period, participatory uses of GIS in France and Belgium remained too rare to constitute a significant field of study (Peribois 2008). Critical GIS analyses remain relatively limited in France (Joliveau et al. 2013; Noucher 2015, 2023; Gautreau 2021; Sahraoui 2025), but several volumes of the ISTE/Wiley

Encyclopedia of Science include chapters that refer to it (Debarbieux and Hirt 2022; Mericskay 2024; Giraut and Houssay-Holzschuch 2023). The CNRS summer school, devoted to the subject in June 2023, which brought together roughly fourty French participants, also attests to its gradual and cautious, yet genuine, entry into the French academic agenda.

C.3. What kind of geospatial world are we making?

Critical GIS already offers multiple answers to the Winner's question. However, we must reflect further on the collective dimension of this *we*, which encompasses not only producers but all those who contribute, consciously or not, to making these geospatial technologies function. This includes computer scientists, engineers and modelers, but also entrepreneurs, investors and policymakers who enable such technologies to materialize, and the countless users, whether they are citizens, scientists, community organizers, public officials, journalists, etc., whose everyday interactions contribute to shaping these systems. Early critical researchers identified the global nature of GIS when they examined the rise of the "information society" (Sheppard 1995; Sheppard et al. 1999). In his most recent book, Wilson (2017) similarly situates his analysis within the pervasive landscape of the location-aware society.

This focus on locational issues in contemporary society is reflected in the proliferation and increasing sophistication of geospatial technologies. Paradoxically, because these technologies are automated and embedded in everyday objects, the sleepwalking citizen remains largely unaware of them. Precisely at a moment when geographic information is merging with information systems and when the spatialization of data extends well beyond territorially framed approaches, it is essential to remain attentive to these informational artifacts. Though elusive, they increasingly carry significant social, spatial and political weight. These technologies assign new roles to geographical space in regulating individual behavior and organizing social interactions. They generate or perpetuate sociospatial inequalities. To restore consistency to analyses of these diffuse forms of technology, it is imperative to reassert the political status of these representational and operational systems, which have become both mundane and unconscious. We must examine how these technical artifacts encapsulate worldviews, embed politically oriented choices, and, implicitly or not, convey values and ideologies that can evolve between their design and their uses. Particular attention must be paid to the social and spatial inequalities they underlie, which remain insufficiently addressed in current research.

Much of this critical analysis of geospatial techniques is carried out by social scientists from various disciplines, including anthropologists, sociologists, political scientists and scholars in information and communication sciences, who have approached the digital issue through the questions and methods of their respective disciplines. Geographers who study issues related to digital society consider space to be central to their research. They have an important role to play in this field. However, because "consumer" tools conceal the geospatial technologies that support them, it is all the more essential for geographers to take them seriously. Meeting this challenge requires strengthening ties with geomaticians who, through their work and interests, maintain continuous dialogue with computer scientists and data specialists. This invites reflection on the issue of "technopositionality" identified by Elwood et al. (2011) and Wilson (2017). How can we combine technical and critical practices in GIS? How can we simultaneously use and interrogate geospatial technologies? Although these questions arise at the individual level for both researchers and practitioners, they can only be answered collectively. In our view, this requires reinforcing the ties between geographic information sciences and geography.

This volume therefore represents a preliminary step toward building a shared culture around the geospatial technical issues to move beyond a purely instrumental conception. The disciplinary alliances at work here, interlinking geography, geomatics, sociology, philosophy and information and communication sciences, lay the foundations for pluralistic approaches capable of deciphering not only the technical reconfigurations, but also the legal, economic, sociological, ethical, political and geographical transformations brought about by the accelerating geodigitalization of the world.

C.4. References

Crampton, J.W. (2010). *Mapping: A Critical Introduction to Cartography and GIS*. Wiley, Hoboken, NJ.

Debarbieux, B. and Hirt I. (2022). *The Politics of Mapping*. ISTE Ltd, London and Wiley, Hoboken, NJ.

Ellul, J. (1977). *Le système technicien*. Le Cherche-Midi, Paris.

Elwood, S. (2022). Toward a fourth generation critical GIS: Extraordinary politics. *ACME: An International Journal for Critical Geographies*, 21(4), 436–447.

Elwood, S., Schuurman, N., Wilson, M.W. (2011). Critical GIS. In *The SAGE Handbook of GIS and Society*, Nyerges, T., Couclelis, H., McMaster, R. (eds). SAGE Publications Ltd, Los Angeles.

Gautreau, P. (2021). *La Pachamama en bases de données : géographie politique de l'information environnementale contemporaine*. Éditions de l'IHEAL, Paris [Online]. Available at: http://books.openedition.org/iheal/9362.

Giraut, F. and Houssay-Holzschuch, M. (2023). *The Politics of Place Naming*. ISTE Ltd, London and Wiley, Hoboken, NJ.

Goodchild, M.F. (2014). Two decades on: Critical GIScience since 1993. *The Canadian Geographer/Le Géographe Canadien*, 59(1), 3–11. doi: 10.1111/cag.12117.

Jacob, C. (2006). *The Sovereign Map: Theoretical Approaches in Cartography Throughout History*. University of Chicago Press, Chicago.

Joliveau, T. (2004). Géomatique et gestion environnementale du territoire. Recherches sur un usage géographique des SIG. Thesis, Université de Rouen [Online]. Available at: https://hal.science/tel-04559188v1.

Joliveau, T., Noucher, M., Roche, S. (2013). La cartographie 2.0, vers une approche critique d'un nouveau régime cartographique. *L'Information géographique*, 4(12), 29–46.

Kwan, M.-P. (2002). Feminist visualization: Re-envisioning GIS as a method in feminist geographic research. *Annals of the Association of American Geographers*, 92(4), 645–661. doi: 10.1111/1467-8306.00309.

Lussault, M. (2009). *De la lutte des classes à la lutte des places*. Grasset, Paris.

Mahmoudi, D. (2020). Critical geographic information system. In *International Encyclopedia of Human Geography*, Kobayashi, A. (ed.). Elsevier, Thousand Oaks, CA. doi: 10.1016/B978-0-08-102295-5.10530-X.

Mahmoudi, D. and Shelton, T. (2022). Doing critical GIS. *ACME: An International Journal for Critical Geographies*, 21(4), 327–336.

Moricskay, B. (2024). *Cartographic Communication*. ISTE Ltd, London and Wiley, Hoboken, NJ.

Noucher, M. (2015). De la trace à la carte et de la carte à la trace : pour une approche critique des nouvelles sources de fabrique cartographique. In *Traces numériques et territoires*, Romele, A. and Severo, M. (eds). Presses des Mines, Paris [Online]. Available at: http://books.openedition.org/pressesmines/2006.

Noucher, M. (2018). Que reste-t-il de Friday Harbor ? Pour une approche critique renouvelée des usages du géoweb fondée sur l'analyse des traces numériques. *Revue Internationale de Géomatique*, 28(1), 15–37. doi: 10.3166/rig.2017.00036.

Noucher, M. (2023). *Blancs des cartes et boîtes noires algorithmiques*. CNRS Éditions, Paris.

Pavlovskaya, M. (2006). Theorizing with GIS: A tool for critical geographies? *Environment and Planning A*, 38(11), 2003–2020. doi: 10.1068/a37326.

Pavlovskaya, M. (2018). Critical GIS as a tool for social transformation. *Canadian Geographies/Géographies Canadiennes*, 62(1), 40–54. doi: 10.1111/cag.12438.

Pavlovskaya, M. and Martin, K.S. (2007). Feminism and geographic information systems: From a missing object to a mapping subject. *Geography Compass*, 1(3), 583–606. doi: 10.1111/j.1749-8198.2007.00028.x.

Peribois, C. (2008). Usages de l'information géographique dans la gestion participative du territoire. Regards croisés en France, en Belgique et au Québec. Thesis, Université d'Angers, Centre Angevin de Recherche sur les Territoires et l'Aménagement (CARTA).

Sahraoui, Y. (2025). Paysages hybrides. Pour une modélisation critique et participative. Volume 1. Volume scientifique et inédit. Thesis, Université Paris 1 Panthéon-Sorbonne.

Schuurman, N. (2000). Trouble in the heartland: GIS and its critics in the 1990s. *Progress in Human Geography*, 24(4), 569–590.

Schuurman, N. (2006). Formalization matters: Critical GIS and ontology research. *Annals of the Association of American Geographers*, 96(4), 726–739.

Sheppard, E. (1995). GIS and society: Towards a research agenda. *Cartography and Geographic Information Science*, 22(1), 5–16. doi: 10/drkfdg.

Sheppard, E., Couclelis, H., Graham, S., Harrington, J.W., Onsrud, H. (1999). Geographies of the information society. *International Journal of Geographical Information Science*, 13(8), 797–823. doi: 10.1080/136588199241021.

Simondon, G. (1958). *Du mode d'existence des objets techniques*. Aubier, Paris.

Staszak, J.-F. (2001). La géographie. In *Epistémologie des sciences sociales*, Berthelot, J.-M. (ed). PUF, Paris.

Thatcher, J., Bergmann, L., Ricker, B., Rose-Redwood, R., O'Sullivan, D., Barnes, T.J., Barnesmoore, L.R., Beltz Imaoka, L., Burns, R., Cinnamon, J. et al. (2016). Revisiting critical GIS. *Environment and Planning A: Economy and Space*, 48(5), 815–824. doi: 10.1177/0308518X15622208.

Wilson, M.W. (2017). *New Lines: Critical GIS and the Trouble of the Map*. University of Minnesota Press, Minneapolis.

Winner, L. (1983). Technologies as forms of life. In *Epistemology, Methodology and the Social Sciences*, Cohen, R.S. and Wartofsky, M.W. (eds). Springer, Dordrecht.

List of Authors

Colette CAUVIN-REYMOND
CNRS
UMR LIVE
Strasbourg
France

Juliette DAVRET
Maynooth University Social Sciences Institute
National University of Ireland
Ireland

Allan DENEUVILLE
UR MICA
Université Bordeaux Montaigne
France

Grégoire FEYT
UMR PACTE
Université Grenoble Alpes
France

Baptiste HAUTDIDIER
UR ETTIS
Inrae Bordeaux
France

Thierry JOLIVEAU
UMR Environnement-Ville-Société
Université Jean Monnet de Saint-Etienne
France

Boris MERICSKAY
UMR Espaces et Sociétés (ESO–Rennes)
Université Rennes
France

Juliette MOREL
Lab'Urba
Université Paris-Est Créteil
France

Matthieu NOUCHER
CNRS
UMR PASSAGES
Pessac
France

Teriitutea QUESNOT
UMR ESPACE
Université Côte d'Azur Nice
Institut Universitaire de France Paris
France

Index

A, C

D, G

H, I, M

N, O

Q, S

T, V

Printed and bound by CPI Group (UK) Ltd, Croydon, CR0 4YY
13/09/2026
14961010-0002